Immersed in Technology

Immersed in Technology

Art and Virtual Environments

edited by Mary Anne Moser
with Douglas MacLeod
for the Banff Centre for the Arts

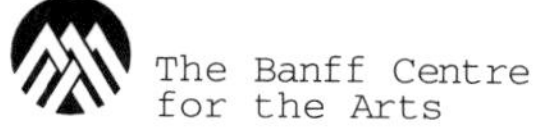

The MIT Press
Cambridge, Massachusetts
London, England

709 04 MOS
Ref 7184

© 1996 Massachusetts Institute of Technology

All rights reserved. No part of this book may be reproduced in any form by any electronic or mechanical means (including photocopying, recording, or information storage and retrieval) without permission in writing from the publisher.

This book was set in Bembo by Compset, Inc. and was printed and bound in the United States of America.

Library of Congress Cataloging-in-Publication Data

Immersed in technology : art and virtual environments / edited by Mary Anne Moser with Douglas MacLeod for the Banff Centre for the Arts.
p. cm.
Includes bibliographical references and index.
ISBN 0-262-13314-8 (alk. paper)
1. Art and technology. 2. Virtual reality. I. Moser, Mary Anne. II. MacLeod, Douglas. III. Banff Centre for the Arts.
N72.T4I48 1995
701′.05—dc20 95-385
CIP

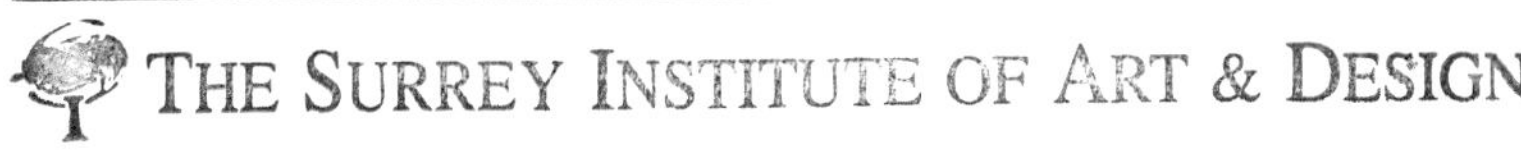
THE LIBRARY
THE SURREY INSTITUTE OF ART & DESIGN

Contents

Preface

A new medium can suggest a multitude of approaches. In 1929, Dziga Vertov's *The Man with the Movie Camera* cataloged possibilities for the evolution of film. From narrative structures to special effects, it shows what cinema could have become. Virtual reality occupies a similar historical moment—it is unformed and hence its possibilities seem unconstrained.

The Art and Virtual Environments Project conducted at the Banff Centre for the Arts in Banff, Canada, was an attempt to insert cultural practice into this realm of possibilities. Although the artistic community has often been excluded from the development of new technologies, this situation is changing. Artists no longer sit on the sidelines eventually to become grateful users of borrowed tools but have become active in development, creating a disturbance in the field with new contingencies.

By providing an opportunity for artists to develop and work with emerging technologies, the Art and Virtual Environment Project was also about the mutation of cultural practice. A new medium like virtual reality challenges traditional conventions not because the participant wears a helmet or glove but because it suggests new relationships between the viewer and the viewed.

The structure of the project was designed to accommodate these challenges. An open call for proposals was issued toward the end of 1991 inviting all artists or groups of artists to submit ideas for projects. Applications received from all over the world were assessed by Banff Centre personnel and seven external adjudicators with backgrounds in art and technology from across Canada. Artistic merit, technical feasibility, and innovation in the use of the medium were all considered. In March 1992 the Centre began discussions with eight groups of artists. At the same time we continued production on Lawrence Paul Yuxwelup-tun's virtual environment artwork that was initiated during the Bioapparatus residency the previous year. All told during this project, nine major virtual pieces were produced.

A fundamental component of the project was the inclusion of writers and theoreticians who would examine this new medium, and thus the call for proposals also requested submissions for essays and analyses. Several chapters in this book resulted from that process, and, as the project gained definition, other writers were invited to participate. Throughout the Art and Virtual Environments Project we brought writers to the Centre to interact with the artists and their projects in an attempt to create a synergy between theory and practice.

By the end of March 1994, after two years of intensive and ground-breaking work, all of the projects were complete. Eight of the nine projects were demonstrated and almost all of the essays were presented at the Art and Virtual Environments Symposium held in conjunction with the Fourth International Conference on Cyberspace at the Banff Centre in May 1994.

The overriding goal of the project, through the artworks, symposium, and this book, was to generate a discourse that could deal with the issues around virtual reality. Sometimes discussions were generated by cutting lines of computer code; at other times it was through papers and presentations. Virtual reality, in fact, was a term that was abandoned early in the process. Loaded with media hysterics and the stigma of violent videogames, it did not accurately describe what we were trying to accomplish. Instead, real-time, three-dimensional, interactive graphics and sound represent the technical parameters of the work.

Within those parameters, however, chaos reigned. We really had no idea what we were doing. This is a very liberating, if exhausting, method of production. It was like staging nine different operas in two years while at the same time trying to invent the idea of opera. While there are tried and true procedures

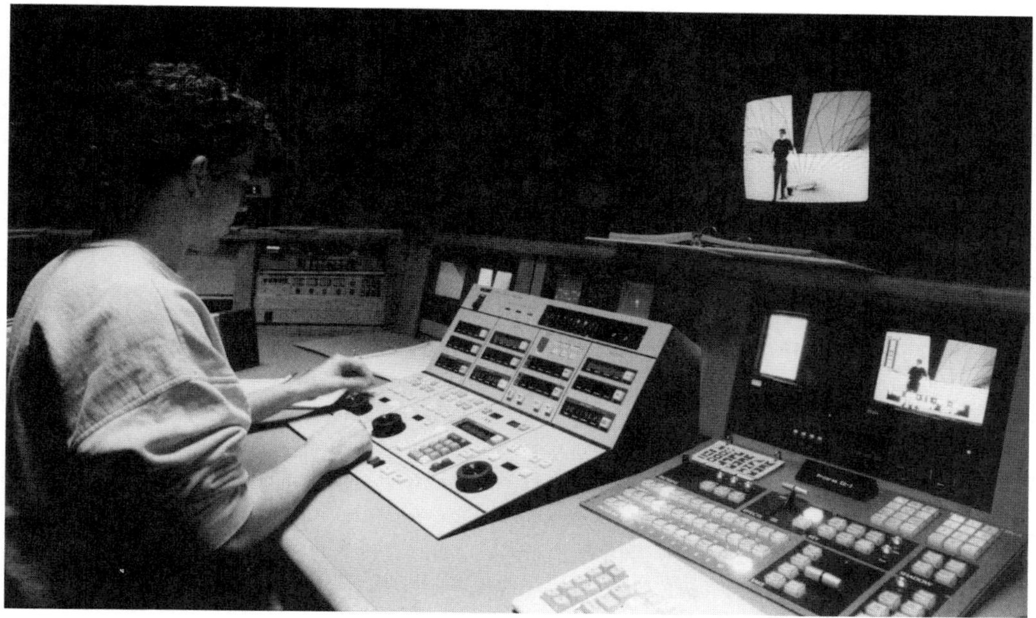

1. Darryl Kaminski prepares a video version of Michael Scroggins and Stewart Dickson's *Topological Slide* (1993). Photo Donald Lee.

for production in theater and film, no such algorithm exists for the creation of a virtual environment. To develop nine major virtual reality projects in two years was only possible because we did not know enough to say it was impossible. But the inchoate nature of the medium also served us well. There was no one to tell us what we could or could not do.

Nonetheless the results are impressive. Toni Dove and Michael Mackenzie's *Archeology of a Mother Tongue* and Will Bauer and Steve Gibson's *Objects of Ritual* describe new genres of performance. Brenda Laurel and Rachel Strickland's exploration of characterization in *Placeholder* is ground-breaking work. Both Perry Hoberman's *Bar Code Hotel* and Ron Kuivila's *VR on $5 a Day* have a manic radiance in their approach to computers and human interaction. Michael Scroggins and Stewart Dickson's *Topological Slide* moves us through one cultural space and Lawrence Paul Yuxweluptun's *Inherent Rights, Vision Rights* moves us through another. Michael Naimark's *Field Recording Studies* continues his seminal work in the recording of actual landscapes. And two versions of *Dancing with the Virtual Dervish* by Diane Gromala, Marcus Novak, and Yacov Sharir pioneered a new approach to multidisciplinary work in cyberspace. The theoretical investigations showed similar innovation. For example, N. Katherine Hayles's analysis of virtuality through a semiotic square and the lyricism of Jeanne Randolph's lists both show the feedback loop of theory and practice. Erkki Huhtamo's study of media archaeology is an invaluable historical document. Each project and essay in its own way makes a valuable contribution to the field.

These ideas and artworks provide only a suggestion of what this medium could be. However, the very frailty of possibilities is cause for concern. Many of these works will never be shown again. Some are simply too complex to remount. In other cases, the team of artists and programmers that produced the piece has dispersed, taking with them detailed knowledge of the assembly and installation of a particular work. Unfortunately, too, as the medium of virtual environments becomes more and more defined, many of these different approaches will be ignored, abandoned, or forgotten as the medium coalesces into a mature form. Most worrisome of all, a new approach to cultural initiatives (not only in Canada but throughout the world) suggests that all explorations must have commercial or revenue-generating potential. While the effects of these pressures remain to be seen, it is very possible that this book documents a project that could never happen again.

In the end, the Art and Virtual Environments Project was not about hardware and software, but rather about the need for a new perceptual structure for the arts. What the projects and the essays in this book demonstrate so clearly is that, as virtual technologies and applications develop, we need to move beyond purely technical considerations in order to actively examine, and adapt to, the changing nature of contemporary artistic practice.

Douglas MacLeod
Project Director

Acknowledgments

Just as artists work closely with technologists in new media, so too do editors rely on the expertise of others. I am grateful to all those who have worked on this project and many more who offered advice along the way.

Daina Augaitis and Michael Century, who first conceived of a book as part of the Art and Virtual Environments Project, have been actively involved in the book's development. Their guidance and efforts in shaping its contents are appreciated and have significantly influenced the resulting publication. Douglas MacLeod sent out the initial call for writers in his role as the director of the Art and Virtual Environments Project and has played an important role ever since, sharing his wit and expertise over the course of the project.

The writers we invited to participate have contributed important essays and been a pleasure to work with. We were fortunate to be able to bring most of them to Banff for the Art and Virtual Environments Symposium held in conjunction with 4Cyberconf. Their energy and ideas (even after three days of presentations) made for a stimulating symposium and, in the pages that follow, an equally fascinating book. With both the symposium and book, we have tried to bring artists and theorists into proximity; we are grateful for their efforts toward the cross-pollination of theory and practice.

The artists have played a central role in this process. They have generously shared the strategies informing their works in their statements in this book, and have described many of the technical challenges. We are grateful for their patience and flexibility in working with this challenging new medium over the past years.

Getting images of the artworks was a more challenging task than initially imagined. Since the entire department staff was involved in the production of the artworks, many of which were up and running for only a few short days, the process of taking pictures and grabbing frames in this time presented more than a few hurdles. Many thanks are extended to all of the staff who made it possible, especially Gordon Tait, Raonull Conover, Marilyn Love, Cathy McGinnis, and Douglas Smith for their extra efforts. Cheryl Bellows and Don Lee did an admirable job of photographing the installations under difficult circumstances. Many thanks are due to the artists again, who did much of the work in preparing images for publication.

Finally, I am grateful to the Banff Centre staff who have made this book possible: Janet Anderson, Deborah Cameron, and Dan Thorburn for administrative help; Mary Squario for assistance in preparing the manuscript; Sara Diamond and Lorne Falk for honest criticism; and the rest of the Media and Visual Arts staff for support in other important ways.

Mary Anne Moser

Introduction

Mary Anne Moser

It seems that the assimilation of technology into everyday life has become unremarkable. One more gadget or another more sensitive tool is easily accommodated by current expectations of applied research. Turning a blind eye to conditions and consequences that make these products available becomes simpler and simpler, as the technologies themselves become more ubiquitous. In many ways, this pattern of assimilation is not new. However, recent developments in digital media have reinvigorated discussions about the cultural impact of technology because the implications of these technologies are, at the moment, unique.

When "virtual reality" was first ballyhooed as the technology to change the world or, better yet, replace it, it was treated as fundamentally different from technologies developed up to that point. This book views it, however, as a technology in a continuum of developments that, on various levels, raise questions more remarkable than the initial allure of the technology itself. As developments continue to race ahead, the essays that follow pause to consider potential consequences. They examine some of the implications—in particular the social, cultural, ethical, and political issues—of virtual environments, also known as artificial reality, virtual space, virtual reality, immersive media, or experiential images.

The common denominator of the above terms concerns an apparent dematerialization of experience. The possibility of programming human presence in immaterial "space" does indeed pose some weighty questions and may explain much of the hype surrounding this new medium. At the same time, this is just another technology and does not change the fact that we continue to operate in the material world. Many pockets of the globe are already immersed in technology, although perhaps not so literally as by virtual environments. Even those that are not immediately surrounded already feel the impact: "exclusion from this virtual realm does not protect economies and cultures throughout the globe from potentially suffering its yet uncertain effects," as Margaret Morse points out in this volume.

To focus our understanding of new media in a cultural context, we long ago abandoned the term "virtual reality" at Banff for the less sensational "virtual environments." Whatever one chooses to call it, the common characteristic of emerging technologies involves the elimination of duration: the collapse of time into real time. While, on a practical level, the instantaneous transmission of information seems quite handy, the interval created by elapsed time provides an absolutely crucial instant to ponder the consequences of the collapse of duration. It serves as a reminder of the dimension of time, which, as Paul Virilio suggests, is under siege by real-time technologies: "They kill 'present' time by isolating its presence here and now for the sake of another commutative space that is no longer composed of our 'concrete presence' in the world, but of a 'discrete telepresence' whose enigma remains forever intact."[1]

To prevent that discrete telepresence from appearing intact, the essays, artworks, and artists' statements brought together here present an unresolved position, not a seamless whole. Taken together, they neither reject nor uphold technological development. Instead, they examine aspects of digital culture and virtuality to understand the interests that drive them forward. In doing so, they reflect on the construction of meaning in and around cyberspace. After all, although it may appear uncharted, the immaterial nonspace of cyberspace is not devoid of meaning. In addition, in this relatively unknown yet value-laden realm, many equivocal positions appear.

One such position underlines the theme of this anthology, which derives from the shift to the experiential from the perceptual reception of the image. (Since this book embarks on a study of cyberspace through the realm of contemporary art, the reference to the image is fitting, although the word "information" could readily be substituted.) As Timothy Druckrey has argued, the bounded image is seen from a distance. It exists unto itself and offers a perceptual experience. Images that implicate the viewer in some way, however, as with interactive or immersive media, are unbounded. They require experiential cognition. The latter put the critical viewer in an untenable position: one must assimilate an image to comprehend it, yet it must also be dismantled in order to reflect upon it.[2]

This condition of both embracing and criticizing our immersion in technology seeds a stimulating debate that, on more than one occasion, has flared vigorously at Banff. The first of these occasions, which was in some ways a precursor to the Art and Virtual Environments Project, was a residency on the Bioapparatus organized by Catherine Richards and Nell Tenhaaf in 1991.[3] In the midst of this ten-week residency for artists, composers, and writers, Richards and Tenhaaf organized a symposium, the Virtual Seminar on the Bioapparatus, during which participants from art and technological backgrounds argued their positions on art and new technologies passionately. While the energy and opposition that characterized those debates has largely been replaced by more collegial interactions, the sense of urgency surrounding these issues has far from abated.

The Bioapparatus residency was significant in the way it highlighted the antinomy that characterized discourse around cyberspace. Proponents of VR were zealous in their defense of no-holds-barred development, envisioning worlds full of possibility once human presence could be better controlled. Critics were equally vociferous: controlled by whom? Accountability for the impact of new technologies on power relations (and thus on social, cultural, and environmental conditions) was demanded. Since then, the hype surrounding what we then called VR has peaked and ebbed. Over time, many of the positions formulated earlier have articulated in more detail the apparently untenable or contradictory aspects of cyberspace, as the essays in this book demonstrate.

The issues addressed in the pages that follow cover a contested territory. For example, the ideological underpinnings that shaped the earliest visions of virtuality contradict the potential it is now seen to hold. Are these two positions

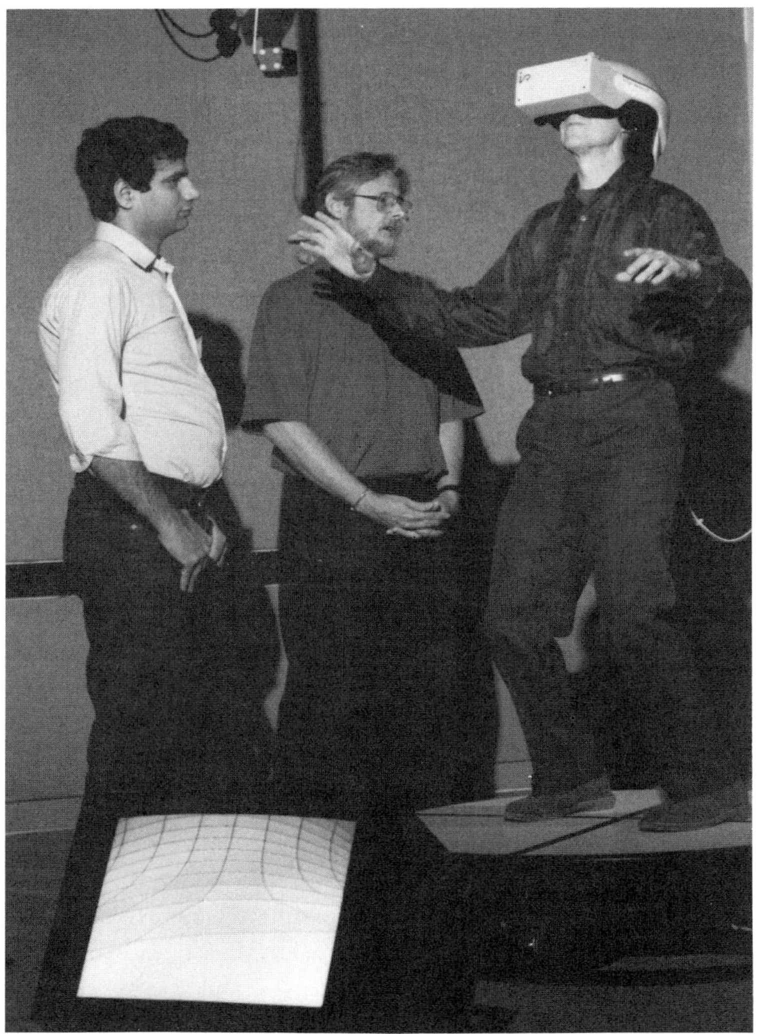

2. Michael Scroggins and Stewart Dickson, *Topological Slide* (1993). Scroggins (center) assists a participant on the slide with a volunteer spotter nearby. Photo Cheryl Bellows.

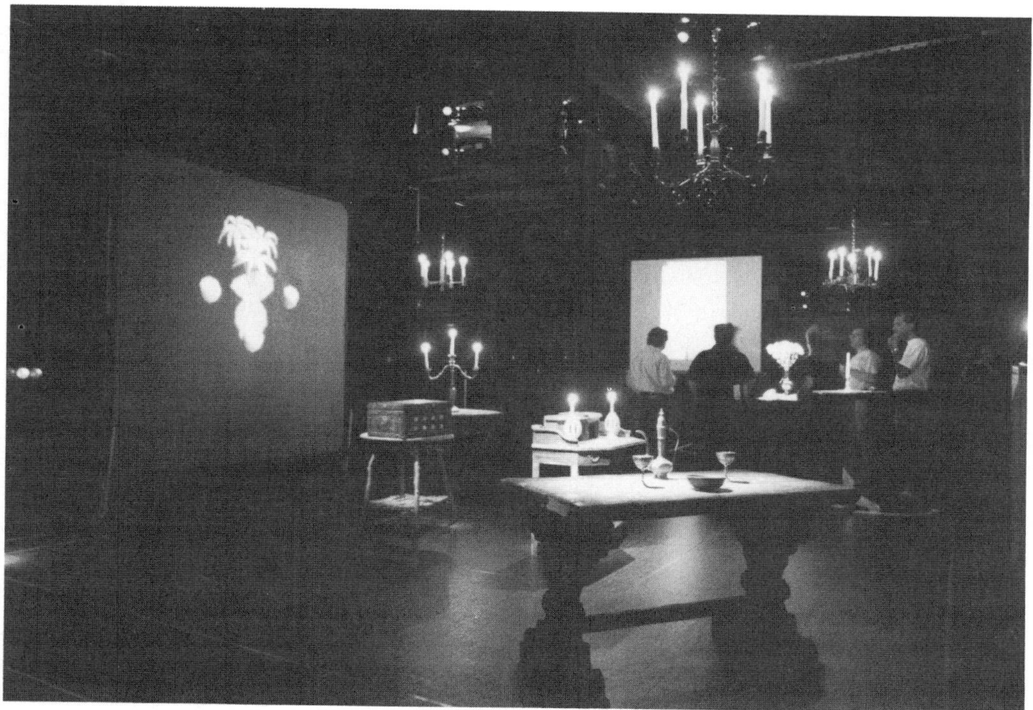

3. Will Bauer and Steve Gibson, *Objects of Ritual* (1994), installation view showing participants at the starting point of the interactive artwork. Photo Donald Lee.

incompatible? Virtual reality technology has been built on the foundations of cybernetics, described by its founding father Norbert Wiener as the science of communication and control.[4] Cybernetics attempted to map entirely the human sensorium and to create, even in its most anarchic state, an environment that is completely programmed. At the same time, the Internet offers opportunities for emancipatory projects and promises to dismantle the hierarchies of power long embedded in other existing communications systems. Just how racial identity and equitable social relations are in fact realized in cyberspace has not been much discussed, although initial research shows promising indications (Bailey).

Questions of race in cyberspace bring discussions securely back to issues concerning the material body. Are we really disembodied in cyberspace, as so much of the discourse around loss seems to suggest? Or is the virtual body a combination of human and nonhuman components, both flesh and information: "If it is obvious that we can see, hear, feel, and interact with virtual worlds only because we are embodied, why is there so much noise about the perception of cyberspace as a disembodied medium?" (Hayles). How we imagine the integration of virtual and nonvirtual into a sense of self shapes the subjectivity of the cybersubject. Rather than presenting the need for an all-new conception of subjectivity that incorporates a doubled body, both material and ephemeral, the immateriality of cyberspace could be seen as constructing subjectivities with which women have long been accustomed (Tenhaaf).

Binary oppositions that characterize power struggles—such as self/other, black/white, male/female—are often treated as though they might evaporate in the disembodied flux of electronic communications. But, so far, there are few signs that this is the case (Bailey, Stone). There is more evidence that, despite the potential for the construction of new subjectivities, old patterns continue to emerge. Concepts such as embodiment, space, reality, community, and authenticity of experience that are now under examination in relation to virtual environments have already been debated and reexamined in the practice of audiophony (Dyson). The outcome of these debates in relation to cyberspace remains to be seen.

Efforts to alter old patterns in a new medium may meet obstacles in part due to the origins of these technologies. Both virtual reality and the Internet were initially developed for military purposes, and significant resources for development continue to come from that quarter. The masculinist and rationalist biases that characterize most technological research programs are still widely influential (Milthorp). The role of the military continues to have enormous impact in conjunction with the media. The Gulf War in particular has altered perceptions of location, contact, and contamination for millions of television viewers who night after night watched enthralled by a new kind of surgical warfare. The role of telepresence in this war pointed to the ethical questions concerning the absence of community in cyberspace: "Can there be an atopicality of the community that nonetheless gathers, a community going nowhere, but ecstatic, a community of shattered egos, where the control towers come tumbling down, and where the other is genuinely anticipated?" (Ronell).

Technological development often races ahead of any reflection on it, as the twentieth century as a whole seems to testify. Consider the fascistic efforts of commercial interests to create a widespread desire for new technologies. Perhaps the unrelenting pace of developments driven by the "technological ethos" can be understood using psychoanalytic theory (Randolph). What is the allure of cyberspace anyway? And at what cost do we assimilate the prevailing images and narratives around virtuality? The environmental impact of the industries of simulation themselves must be factored into the equation. Aboriginal approaches to cyberspace may yield insight into notions of agency that are more responsible to the material environment from which there is no escape (Todd).

Although new imaging technologies are now being developed in parallel by military, commercial, and artistic interests, different motives may be driving the research in these domains. Artists' interests are complicated by the fact that they remain largely reliant on corporations and institutions for access to the technology. Their motives in it may be poles apart from those of corporations yet, in order to get access, they have to offer the corporation something in return. Is there a way that artists can work with this medium without heavy constraints imposed by industry? As Douglas MacLeod points out in his preface to this book, support for artists using new technologies is increasingly linked to the revenue-generating or commercial potential of their proposed projects. On the other hand, artists have a history with images, and are able to construct meaning in representational systems based on considerable experience with the social relevance of image and perception. This is the power that lies with artists, which is ever-increasing as the image acquires new currency in mass culture. Artists working in this domain are renegotiating their positions through constantly reworked aesthetic strategies that relate to the art world and beyond (Huhtamo).

What is the artist's role precisely and how do aesthetic features of cyberspace relate to the viewer's experience and conception of the work? Who has access to the technology and how will they apply their power? "It is then a task of art to give shape to possibilities and questions about this socioeconomic shift and its emerging cultural forms that are not raised by virtual environments produced and designed for instrumental purposes or for entertainment" (Morse).

Virtual environment artworks and the issues they raise provide a challenging and stimulating point of entry to cyberspace for many of the essays in this book. Through its support of new media artworks, the Banff Centre continues to embrace both the production and examination of technology, considering critical questions about access, interests, and ideology while continuing to develop its strategic alliances with industry. This book also attempts to reflect this position: while actively supporting new media technologies and gratefully acknowledging the support of corporate partners, it offers a critical eye on developments. The essays, artworks, and artists' statements that follow present aspects of technology in contemporary art and culture that may be remarkable after all.

Notes

1. Paul Virilio, "The Third Interval: A Critical Transition," in *Rethinking Technologies*, ed. Verena Andermatt Conley (Minneapolis: University of Minnesota Press, 1993), 4.

2. This was pointed out by Timothy Druckrey during "The Transient Image/L'Image Transitoire: A Symposium on the Changing Status of the Image," Banff, Canada, November 4 and 5, 1994.

3. The term "bioapparatus" was coined by Catherine Richards and Nell Tenhaaf.

4. See N. Katherine Hayles, "The Seductions of Cyberspace," in *Rethinking Technologies*, 174.

Immersed in Technology

Embodied Virtuality: Or How to Put Bodies Back into the Picture

N. Katherine Hayles

Cyberspace, we are often told, is a *disembodied* medium. Testimonies to this effect are everywhere, from William Gibson's fictional representation of the "bodiless exultation of cyberspace" to John Parry Barlow's description of his virtual reality (VR) experience as "my everything has been amputated." In a sense, these testimonies are correct; the body remains in front of the screen rather than within it. In another sense, however, they are deeply misleading, for they obscure the crucial role that the body plays in constructing cyberspace. In fact, we are never disembodied. As anyone who designs VR simulations knows, the specificities of our embodiments matter in all kinds of ways, from determining the precise configurations of a VR interface to influencing the speed with which we can read a CRT screen. Far from being left behind when we enter cyberspace, our bodies are no less actively involved in the construction of virtuality than in the construction of real life.

If it is obvious that we can see, hear, feel, and interact with virtual worlds only because we are embodied, why is there so much noise about the perception of cyberspace as a disembodied medium? Perhaps we can find a clue in how the constructions of virtuality are constituted. To create the illusion of disembodiment, it is necessary to draw a sharp boundary between the body and the image that appears on screen, ignoring the technical and sensory interfaces connecting one with another. Then the screen image (which can be made to appear three-dimensional by exploiting certain characteristics of human visual and aural processing) is reified, treated as constituting a world opening up behind the screen, an alternative universe that our subjectivities can inhabit.

The final step is to erase awareness of the very perceptual processes that brought this "world" into being. The tautology that makes the illusion compelling is now close to the surface: we want this alternate world to exist so that the body can be left behind, and we know the body can be left behind because this world exists.

Why do we want to leave the body behind? No one has inscribed this desire more clearly than Hans Moravec, head of the Carnegie-Mellon Mobile Robot Laboratory. In *Mind Children,* Moravec argues that the age of the protein-based life forms is drawing to a close, to be replaced by silicon-based life forms.[1] Humans need not despair, however, because they can have their consciousness downloaded into a computer. In the fantastic scenario in which he imagines this operation, Moravec has a robot surgeon cut away a human brain in a kind of cranial liposuction until all the information the brain contained is inside the computer and the skull is empty of brain tissue. Moravec reasons that once human consciousness is safely ensconced inside a computer, it is effectively immortal. If the computer begins to wear out, consciousness can simply be transferred to a new machine.

In this scenario and throughout his text, it is apparent that Moravec equates subjectivity with the mind. The body is treated as a flawed and unwieldy vehicle, necessary in the early stages of human evolution but now become more trouble than it is worth. In the new age of virtuality, as Moravec sees it, the mind will continue unchanged in its incarnation as electronically coded information, and the body will fade away as a superfluous accessory.

Traditionally the dream of transcending the body to achieve immortality has been expressed through certain kinds of spiritualities. Dust to dust, but the soul ascends to heaven. Moravec's vision represents a remapping of that dream onto cyberspace, with an important difference: reversing a long-standing opposition between science and religion, he enlists technoscience as the ally of out-of-body transcendence. To achieve this apotheosis one does not need spiritual discipline, only a good robot surgeon. Such a vision is nurtured by a cultural tradition that has long dreamed of mind as separate from body.

At the end of the twentieth century, it is evidently still necessary to insist on the obvious: we are embodied creatures. Fantasies asserting otherwise can have dangerous consequences. Although this is not the place to debate the technical feasibility of Moravec's project, it is worth remembering that no technology exists that can come remotely close to transferring mind into computer. Moreover, even if such a transfer were possible, mind inside a computer could not possibly remain unchanged from humanly embodied mind. Human consciousness has evolved through millennia with the specific physical structures that comprise its instantiation. Remove these, and everything else changes as well.[2] More objectionable still, the scenario represents an escapist fantasy that late twentieth-century inhabitants can ill afford. Scratch the surface of Moravec's rhetoric and that of like-minded visionaries, and you will find scarcely concealed anxiety about our continued existence on a planet despoiled by environmental poisons and decimated by AIDS. Ironically, such fantasies may be complicit in furthering the very anxieties that engender them. If we can live in computers, why worry about air pollution or protein-based viruses?

The fantasy also has implications for gender politics. Inscribing the female power of reproduction into a technological scenario of (male) parthenogenesis, it identifies subjectivity with the rational mind that has traditionally been encoded masculine, leaving behind the materiality of the body that has been identified with the feminine. Male form, female *materia;* male seed, female ground.[3] Now the (male) technoscientific mind devises for itself a new body, not born of woman, that it imagines will be more suited for its rational thought processes and immortal yearnings. To unpack the implications of these associations, notice that one set of dualisms, male/female, reinforces and powerfully interacts with another, mind/body. As Nancy Leys Stepan has shown using instances of race and gender, the characteristics of one duality can transfer onto another when the two are consistently associated.[4] She instances the notion, popular in the late nineteenth century, that "women are the blacks of Europe." How do such metaphoric equivalences come about, and what implications do they transport from one discursive register to another? In the black/white duality, the black race is discursively constructed as the opposite of the white race, which is assumed to be primary and originary. Women are similarly constructed as the "opposite" sex to men, who are the original and normative sex. Opposite to blacks are white (men) and to women, (white) men. The analysis reveals the implicit or hidden terms, (white) (men), that are understood but not stated in the white/black and men/women dualities. For the hegemonic terms, *white* and *men,* the hidden components create an area of overlap, white men, around which privilege and value can cluster. For the stigmatized terms, *black* and *women,* the overlap creates an association that brings them into correspondence—hence the expression equating blacks and women. The regulated exchanges that comprise this traffic take place in discourse, but the effects are not limited to verbal formulations. In this instance, they led to extensive studies in physiogomy attempting to show that women and blacks had smaller heads, and therefore supposedly less capable brains, than white men. Here metaphoric equivalences were substantive, for they sparked experiments designed to transmute them into scientific fact.

Similar metaphoric mappings underlie the erasure of the body from cyberspace. The dualities line up as follows: mind is superior to body; silicon technology is superior to protein organism; man is superior to woman. Therefore replace the body born of woman with a computer that can serve as a fitting receptacle for the (male) mind. The privileged terms (mind, computer, male) are linked together in mutually reinforcing connections that seem to make it possible to erase or leave behind the stigmatized terms (body, organism, female). The construction of cyberspace thus happens in two ways simultaneously, through technological interventions and discursive practices. To put the body back into the picture, both need to be addressed.

Cyberspace discourse and technology, metaphor and practice, come together in the concept of virtuality. Let us linger at this crossroads, looking at the high-speed traffic across it as a means of understanding the associations and connections that are in the process of constituting the virtual body. For our purposes, virtuality can be defined as *the perception that material structures are interpenetrated with informational patterns.* To see how virtuality marks a crossing, consider the Smart House designed by the American Homebuilders Association at a cost of six million dollars. The Smart House uses state-of-the-art equipment to tie all the major house systems—lights, heating, plumbing, windows, and so forth—into a central computer. Lights and heating are adjusted automatically when the owners return or leave; doors lock and unlock appropriately; window shades are drawn or opened. Each of these objects exists as a material structure and can be understood as such. The thermostat, for example, consists of a plastic case, a thermometer, various colored wires, and a mercury switch. But the objects can also be understood as components in an information circuit. The thermostat communicates with the central computer, the air conditioner, and the heating system. The thermostat is virtual because it cannot fully be understood as a material structure; only when its communication and control function is considered does its design make sense. This view of the thermostat is hardly new, dating back to its inception in the nineteenth century. What is new is the extension of this view to a larger environment, in and out of the Smart House. Insofar as the Smart House is doubly encoded, existing both as physical objects and as instantiations of information flows, the Smart House is a microcosm of contemporary first world culture.

Nowhere is this trend toward virtuality more apparent than in the construction of the human body. Since the early twentieth century, the body has been understood in a variety of fields both as a physical structure and as an informational pattern. In the 1940s, Erwin Schrödinger urged biologists to see the body as an expression of genetic information in his influential work *What Is Life?,* a challenge that has been taken up in the Human Genome Project.[5] Coming from the different tradition of cybernetics, Gregory Bateson in the 1950s asserted to his contemporaries that "information is primary."[6] What this meant had been laid out by Norbert Wiener's 1948 *Cybernetics,* which modeled complex behavior through electrical circuits employing the cybernetic principles of negative and positive feedback loops.[7] By the 1980s, the body had been conceptualized in such cyberpunk novels as *Neuromancer* as a cumbersome appendage that consciousness could leave behind when it vaulted into cyberspace.[8] By 1990, researchers at Xerox Palo Alto Research Center (or, as it is known in the trade, the PARC) had begun to talk about embodied virtuality, envisioning a virtualized environment that would communicate fully with the human subject through a variety of miniature computers embedded in and on the body.[9] Employees at the PARC are provided with interactive badges that communicate with sensors in the ceilings, which in turn signal information about the employees' locations to a computer network. In this practice the body, understood biologically as an expression of genetic information, communicates with its environment through external and internal information circuits into which it is fully integrated. Like the objects surrounding it in the Smart House, the body is neither simply material object nor informational pattern but both at once.

The crossings and interpenetrations constituting the virtual body call for a more sophisticated and nuanced approach than simple binary thinking can provide. It is when one duality is chosen over another—when the body is seen only as information—that its erasure seems possible. Putting the body back into the picture requires a mode of analysis that can complicate and unravel the simple dualisms that underlie its erasure, while still acknowledging the force and efficacy of these dualisms in creating cultural constructions. For this task, I will use the semiotic square, a technology of discursive analysis developed by A. J. Greimas and transported into the United States by Ronald Schleifer, among others.[10] The semiotic square is designed to unpack the implications inhering in a binary pair by making explicit the hidden terms that help to stabilize meaning and generate significance. As with the white/black and men/women dualities, explicating areas of overlap shows how associations travel across metaphoric networks to do political work within the culture. Once the concealed terms have been made explicit, I will use them to explore more fully how the virtual body is discursively constituted. Then I will turn to the practice of three artists working in virtual reality technologies to show how the simulations they create challenge, in another mode, the assumption that cyberspace is a disembodied medium.

The Semiotic Square and the Virtual Body

The construction of a semiotic square begins with the choice of a binary pair. Since I am concerned with the material presence of the body, the first term I choose is presence. The second term is generated by taking the negative of the first term: absence. Presence and absence thus comprise the primary duality for my square. It is no accident that this duality has played a central role in the development of poststructuralist theory. In deconstructive analysis the plenitude and self-sufficiency of presence is revealed as a linguistic illusion, a sleight-of-hand that texts practice when they bootstrap themselves into existence by positing an origin outside themselves.[11] Deconstruction points out that presence, far from being self-evident, can stabilize its meaning only through its difference from absence. Thus the duality of presence and absence in the semiotic square signifies concepts in dynamic interplay with each other rather than independently existing terms. This interplay is indicated by a double-headed arrow connecting the terms.

The Beginning of a Semiotic Square

Presence ↔ Absence

The second duality needed to transform the figure into a square is generated by choosing a term that is like the privileged first term of presence, but different from it by being absent where it is present. (The language here is confusing since presence and absence have already been put into play, but bear with me. The relationship between the first and second pairs should become clear shortly.) The purpose of choosing the second duality is to unfold implications contained in the first pair. There is thus no unique solution but a series of solutions, each of which unfolds different implications. Since my interest lies in the construction of the body in information technologies, I choose as the third term a concept important to information theory: pattern.[12] Pattern is like presence in that the recognition of pattern is often associated with the perception of presence. When the pattern of a dark blob is recognized, for example, it becomes a tree rather than a shadowy mass. But unlike presence, pattern need not imply the material existence of an object. Films, for example, rely on viewers to recognize visual patterns on the screen, even though the objects figured by these patterns are absent. The fourth term is generated by taking the negative of pattern: randomness. We are now in a position to lay out the four terms of the square.

The Complication of a Duality in the Semiotic Square

materiality
Presence ↔ Absence

information
Randomness ↔ Pattern

Since the interplay between presence and absence generates the material world, the axis connecting these terms is labeled "materiality." Similarly, the interplay between pattern and randomness generates "information." Information is formally defined by the theorems and assumptions of information theory, formulated by Claude Shannon in the late 1940s.[13] In its technical sense, information remains conceptually distinct from markers that carry it, such as radio waves or newsprint. The quantity of information in a message is mathematically defined as a function of the probability distributions of the message elements. The distributions are calculated along a scale that ranges from complete predictability to complete unpredictability or randomness. In his commentary on Shannon's theory, Warren Weaver explained that information must have some pattern to distinguish it from noise. If it is entirely predictable, however, it is mere repetition rather than new information. A message conveys maximum information when it is balanced between surprise and expectation. Thus information evolves as a dance between pattern and randomness. The lower axis expresses this interplay.

Now that both sets of dualities are in place, the semiotic square can be used to investigate the implications of virtuality as a crossing between materiality and information. These implications are made explicit by considering the relationships that tie the various terms to one another. The semiotic square is designed to exhaust the relational possibilities between the various terms. For our square, the diagonal connecting presence with pattern can be labeled "replication," for when presence and pattern coincide, object and form are united with no dissonance or separation between them. This is the realm of mimesis, ruled by commonsense assumptions about objects that maintain their form through time. The diagonal connecting the negative terms of absence and randomness can be labeled "disruption." Just as absence disrupts the plenitude of presence, so randomness disrupts the expectation that pattern will be replicated over time.

Virtuality and the Semiotic Square

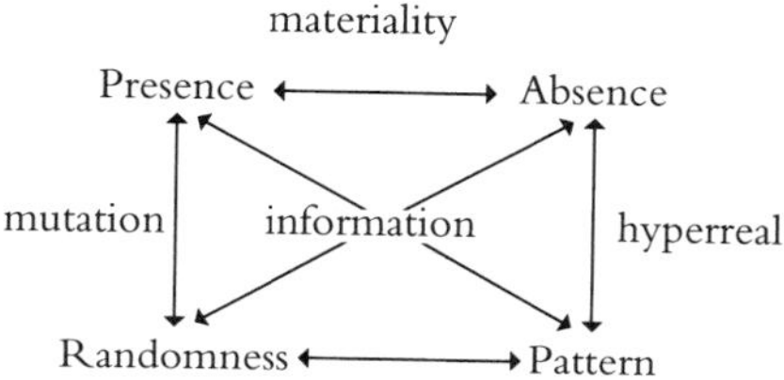

The vertical axis connecting absence and pattern signifies what Jean Baudrillard has called the "hyperreal."[14] The hyperreal comes about, Baudrillard explains, when the chain of displacements connecting a series of imitations to an original becomes so attenuated that the original is lost as a referent. Consider, for example, an audio compact disk produced by sampling sounds from several different recording sessions and digitally manipulating them to achieve the desired effects. There never was a recording session that sounded like the music on the disk, yet clearly the disk is a copy of something. In fact it is a simulacrum, a copy without an original. The simulacrum gives a sense of being a copy because it replicates a pattern, but in the absence of a referent it becomes a copy without an original. The phenomenon of the hyperreal is predicted by the semiotic square as a result of the interplay between the axes of materiality and information.

Continuing with our analysis, we can label the vertical axis connecting presence and randomness as "mutation." Mutation occurs when random variation becomes physically manifest in a material object. Random variations in the genetic code, for example, can result in a freak or mutant. It is no accident that mutation has become an important theme in postmodern literature and culture, for when information is the coin of the realm, the dance between pattern and randomness is central.[15] The mutant is visible proof that randomness is always already present, capable of disrupting pattern at any moment. As with the hyperreal, the semiotic square predicts that mutation would become an important topic once presence/absence is put into interplay with pattern/randomness.

As numerous commentators have pointed out, including Fredric Jameson and Shoshana Felman, the four nodes of the semiotic square recall the four quadrants of a Cartesian graph.[16] The Cartesian analogy helps to explain why the positive term of the second pair, pattern, is placed on the lower right rather than the lower left. In Cartesian grids, the lower right quadrant represents a positive x-value combined with a negative y-value. Recall that pattern is generated by taking the absence of the positive first term, presence; it therefore represents the combination of a positive and negative, analogous to the third quadrant. The fourth term, randomness, is produced by taking the negative of the third term, pattern, which is already marked by negativity. Thus the fourth term represents a negation of a negation. Because of this double negation, it is the least explicitly specified of all the four terms, and therefore the most productive of new complications and insights. It is from the "elusive negativity" of the fourth term, as Felman calls it, that the new is likely to emerge, for the fourth term carries within it the most noise and the fewest expectations about what it will be.

What does the semiotic square as developed above tell us about the virtual body? It shows schematically the possible relationships that can emerge when materiality and information mutually imply each other, thus providing a theoretical framework in which such apparently diverse ideas as hyperreality and mutation can be understood as different manifestations of the same underlying phenomenon. In David Cronenberg's *The Fly,* for example, a scientist believes that if he can transform his body into information, he can transport it instantaneously across distance. He tries the experiment on himself, but a fly inadvertently enters the chamber with him, so that the information of the two bodies gets jumbled together. Is the body that comes out of the transformer after this event the original, or a copy? We might be tempted to say it is the original—at first the scientist appears to be exactly what he was—until the horrifying changes start that will transform him into a monster. As these differences erupt into physical expression, it becomes clear that a random event has intervened to alter the pattern of the man's genetic information.

So the body that emerges is not the original but an imperfect copy. But then, the putative original of this copy bears within itself the stamp of an earlier event (the cell division and chromosome mixing that occurred when sperm fertilized egg) that marks it as a copy rather than an origin. The information contained in that sperm and egg also came from prior events, namely the fertilizations that created the phenotypes who produced the sperm and egg. Thus the origin is displaced into a receding series of genetic events stretching back into the dim mists of prehistory. This displacement provides the basis for Richard Dawkins's argument that human beings are mere "lumbering robots" operated by our genes, who manipulate us for their own selfish ends.[17] Dawkins almost certainly had no knowledge of Baudrillard's work when he wrote *The Selfish Gene,* but it is not difficult to see that, in Baudrillard's terms, his argument implies that human beings are simulacra, copies without originals. The power and efficacy of being associated with an origin is taken out of the realm of human subjectivity altogether and given to the "selfish genes." When the body is an informational pattern, the origin is understood as the point where the pattern begins, not where spirit or breath enter. Like the mutant, the selfish gene testifies that the solidity of flesh is interpenetrated and rendered vulnerable by the informational patterns that determine its destiny.

The devastating effect this interplay between pattern and presence can have on traditional concepts of identity appears in a different mode in *The Fly,* which teaches that the physical durability of the body is an illusion. When the body is an expression of an informational pattern, randomness can always intervene to disrupt or change that pattern. Moreover, although the eruption of randomness into the body is represented as an accident in the film, it is not really an accident but an inevitability waiting to find expression in the physical world, for the very existence of pattern implies a background of nonpattern or randomness against which pattern can be perceived as such. Because pattern and randomness do not exist in isolation from one another, it is impossible to guarantee that one will not leak into the other. I am reminded of the wonderfully oxymoronic phrase that Emerson used in a euphemism for death: "If perchance the inevitable should happen . . ." The disruption of pattern by randomness is as chancy—and as inevitable—as death.

The semiotic square, evoking a space in which randomness can couple with presence and absence can interpenetrate pattern, suggests that the evolution of deconstruction is not yet complete. Beyond poststructuralism lies the posthuman, which can be understood as the realizations that await us when the dialectic of presence/absence is integrated with the dialectic of pattern/randomness.[18] Put another way, the posthuman represents the construction of the body as part of an integrated information/material circuit that includes human and nonhuman components, silicon chips as well as organic tissue, bits of information as well as bits of flesh and bone. The virtual body partakes both of the ephemerality of information and the solidity of physicality or, depending on one's viewpoint, the solidity of information and the ephemerality of flesh. I believe that those who nostalgically try to deny the importance of information in understanding the virtual body are as misguided as those who, like Moravec, try to reduce the body only to information. Either position misses the complexities of the crossings and interpenetrations that constitute the virtual body at this cultural moment.

Crossing the Virtual Divide

The cultural constructions that identify masculine subjectivity with the mind, female subjectivity with embodiment, may be responsible for actual physical differences in how men and women react to immersion in VR. When a user

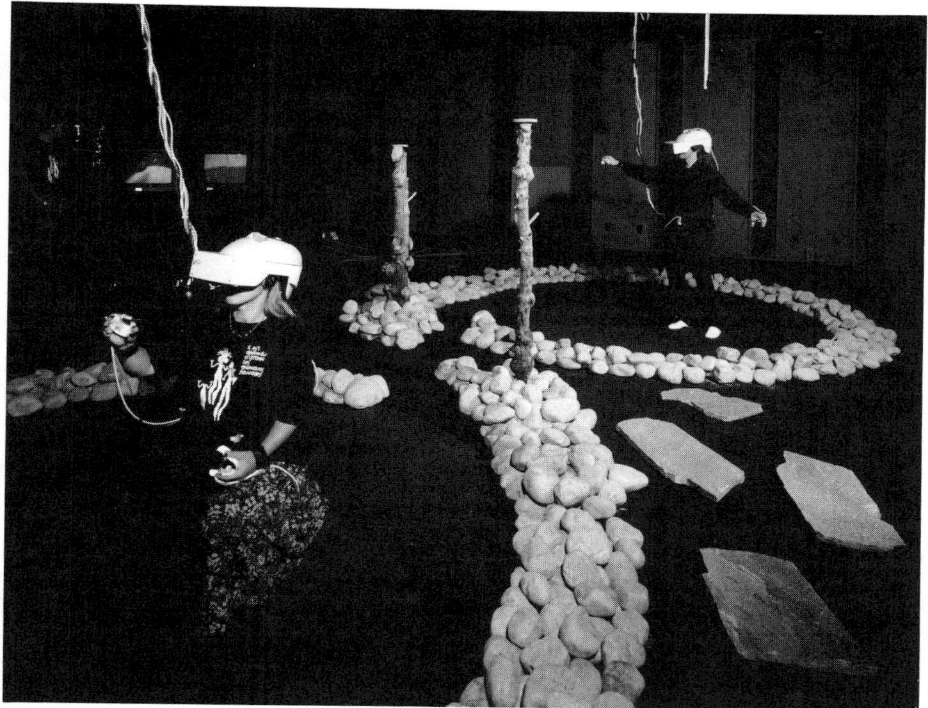

4. Brenda Laurel and Rachel Strickland, *Placeholder* (1993), installation view with programmers Dorota Blaszczak (foreground) and John Harrison in the helmets and Glen Fraser (far left) watching the monitors. Photo Donald Lee and Cheryl Bellows.

enters a VR simulation, body boundaries become ambiguous. Body motions affect what happens in the simulation, so that one both is and is not present in the body and in the simulation. The body marks one kind of presence; the point of view, or pov, that constructs the user's position within the simulation marks another. As a marker of subjectivity, pov is more than an acronym, more even than a noun. In the parlance of VR, it functions like a pronoun, a semiotic container for subjectivity. According to anecdotal evidence from a number of researchers, including Sandy Stone and Brenda Laurel, women are apt to feel more disoriented by the transition to and from VR than men, and they are more prone to motion sickness while in VR.[19] The reason, these researchers conjecture, is that men adapt more readily to the idea that pov can move independently of the body, whereas women are accustomed to identifying pov with the body. Although this informal observation remains to be tested by a systematic study, its performance is already being staged in a variety of cultural documents.

It forms a powerful subtext for *Neuromancer,* the novel that made cyberspace a household word. Case, the male protagonist, sees his body as so much "meat." He feels completely alive only when he is in cyberspace, where his physicality is transformed into a dance of information. Molly, assigned to protect Case and to run interference for his hacking, is a cyborg warrior who delights in her physical prowess. Whereas virtuality for Case means leaving his body behind, for Molly it means modifying her physicality to make it into a more effective weapon. Occasionally Case will need sensory input when he is in cyberspace. By flipping the appropriate switch, he can transport his consciousness inside Molly's sensorium and voyeuristically share her sensations of her body. The difference between riding in a sensorium as a passenger and being constituted through the sensorium is sharply drawn when Molly is injured. Whereas Case can flip back to cyberspace to escape the pain, Molly has no choice but to endure it. Gibson's text illustrates how the experience of virtuality gets overlaid with an encoding that constructs the ephemerality of information as masculine and the materiality of the body as feminine. Virtual reality is no exception to the observation made by many feminist theorists that wherever dualistic hierarchies exist, the privileged term is identified with masculinity and the stigmatized term with femininity.

Feminist responses to a construction of cyberspace as an escape from the body are enacted along a spectrum of resistance, from contestations of what physicality means to reinterpretations of what it implies to reconfigure the physical body with virtual stimuli. Each of these interpretations is struggling to establish itself in a field dominated by militaristic values and male high-tech culture; none is secure from reappropriation by masculinist projects. Although the artistic projects I want to discuss operate from different assumptions, they share a common goal of forcing a reexamination of physicality in an age of virtuality. Moreover, even an analysis of their differences helps to vivify the stakes in how the technology is interpreted and understood.

Brenda Laurel and Rachel Strickland's *Placeholder* simulation illustrates one kind of artistic response.[20] Laurel and Strickland wanted to create a simulation that would model for the VR industry a different aesthetic than the violence, sexism, and racism that are staples of video game productions. They envisioned a simulation that would situate embodied actions within an interactive environment. They wanted the aesthetic choices to grow out of a respect for the complex interactions that take place between embodied creatures and the world they inhabit. Reflecting these priorities, the first choices they made were the sites that would serve as models for the environments recreated in the simulation. They explored the breathtaking terrain around Banff and finally settled on three locales: the vertical rock formations known as the hoodoos, revered by native people as the home of spirits; a natural cave; and a mountain waterfall. They shot panoramic videotapes of each site at four different times of the day and used the tapes to lay video tiles onto wireframe models to create the landscapes for the simulation.

Having chosen the sites, they worked with actors from Precipice Theatre, an improvisational troupe, to create narrative material that could be incorporated into the simulation. Some of this work took place in the troupe's regular rehearsal space, but some of it was done on location. The actors hooted at the hoodoos, splashed in the waterfall, dripped in the cave along with the stalactites.

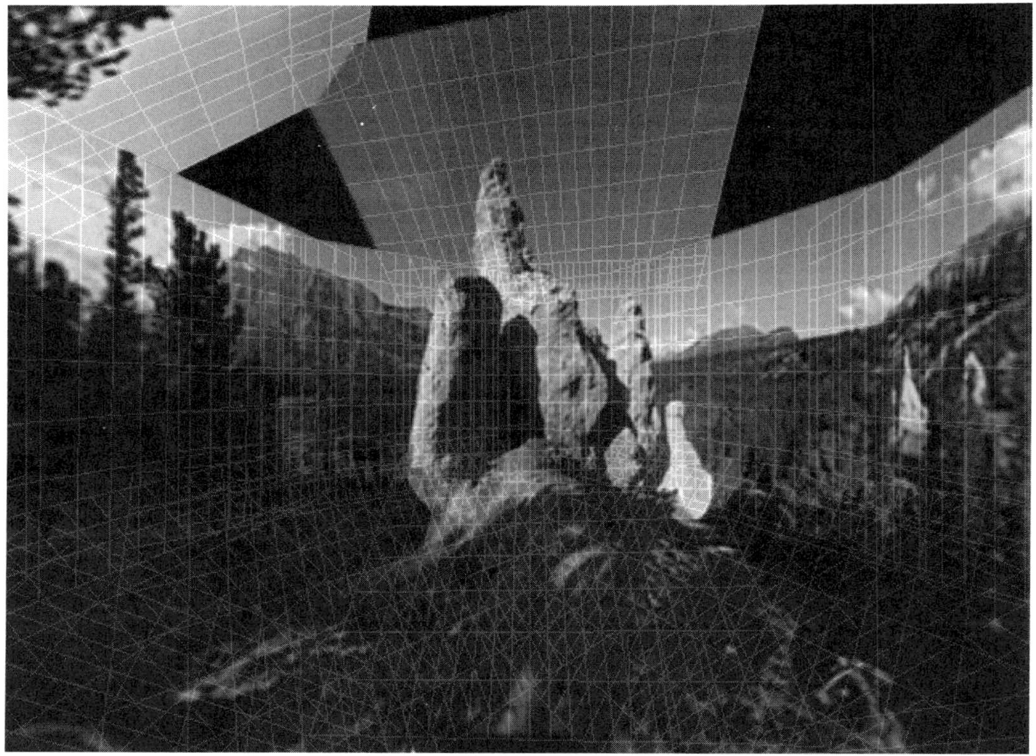

5. Brenda Laurel and Rachel Strickland, *Placeholder* (1993), video image showing the tiling for the Hoodoos environment. Digital image courtesy the artists.

As they interacted with the environments, the idea grew that the simulated environments should be animated. Laurel and Strickland commissioned Russel Zeidler, an architect, to come up with an icon, reminiscent of a petroglyph, that could be used to give a local habitation and a name to the simulated landscape. Zeidler created a schematic face that functions as a "voiceholder," a virtual container that holds recorded messages. When a user touches the face icon, its eyes open. The user can then collect the recorded messages that others have left or leave a calling card herself.

Embodiment in *Placeholder* is enacted through Smart Costumes. To interact with the simulation, a user chooses one of four forms: spider, crow, snake, or fish. In the simulation, the user visually appears as her chosen form. Moreover, voice filters are used to alter the user's voice in ways characteristic of the four Smart Costumes. Crow sounds raucous and masculine, spider wise and feminine, whereas snake and fish are gender-indeterminate. Movement functionalities are determined by the form of embodiment the user chooses. Crow can fly, for example, while the other forms cannot. With snake, vision is affected as well, shifting into the infrared. It is important to note that the simulation does not restore a "natural" connection between the user's pov and the body, most obviously because human beings do not naturally have sensoriums that process information in these ways. The person who experiences this simulation is a techno-bio-subject whose body has been resurfaced and reconfigured by its interface with the technology. What the simulation does insist upon, however, is the connection between pov and incorporation. It is not possible, the semiotics implies, to have experiences without a viewpoint constituted through sensory-motor apparatus specific to the form pov occupies.

Designed to foster interaction with the environment, *Placeholder* encourages interaction between participants as well. The simulation can accommodate two users at once. They stand in neighboring circles formed of river stones. As they interact with the simulation, they move about, learning the limits of the cables connecting them to the simulation through the tactile feedback of stubbing their toes on the stones. Each can see the other in the simulation and hear the voice-filtered comments her or his companion makes. Participants can shapeshift

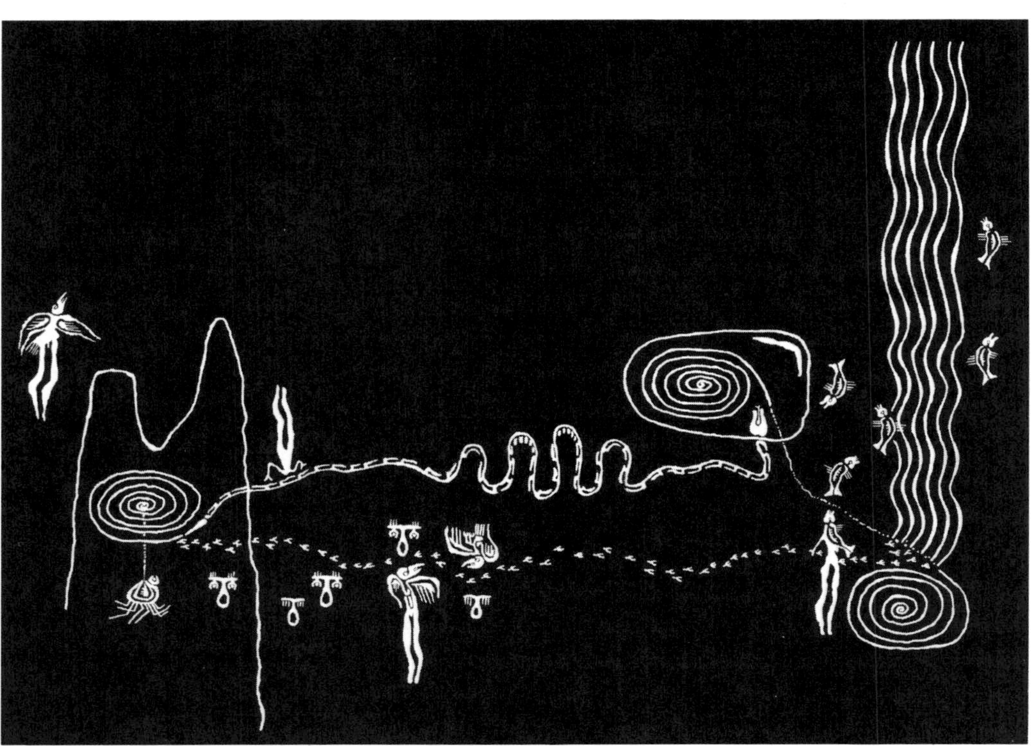

6. Brenda Laurel and Rachel Strickland, *Placeholder* (1993), map. Digital image courtesy the artists.

by touching the appropriate totemic icon. Improvising on cues provided by the environment and each other, they create narrative. To further enrich the narrative possibilities, the simulation provides for a Goddess, a role played by someone who watches the simulation on monitors and interjects comments, suggestions, and commands. The Goddess's voice is spatially localized over the user's head, giving the eerie sensation of coming from *inside* rather than outside. This role was often played by Laurel, and occasionally by others.

A simulation this elaborate does not come cheap. Consisting of over 25,000 lines of code, it ran on no less than eleven computers, including three Onxy Reality Engines and a Macintosh Powerbook. Even so, the simulation remains unfinished, only a partial realization of the ambitious project that Laurel and Strickland had intended. Plans called for further development of the narrative material provided by the Precipice Theatre group, amplification of the sensory modalities associated with the different totemic forms, and many more runs with different kinds of participants, along with even more interviews afterward to get their reactions to the simulation and their interpretation of the experience. Despite the stated aim of providing an alternative for the video game industry, the simulation is not commercially viable in its present form. It requires such a massive amount of equipment and provides so many opportunities for breakdowns that in fact it never got beyond the prototype stage. Most of the runs of the full simulation occurred at Banff during an intense but relatively short period. Since then the machines that ran it have been taken off that assignment and used elsewhere. At present the simulation exists less as a running program than as the videotape records that Laurel and Strickland made to document its operation.

From these records, Laurel and Strickland produced a video on the making of *Placeholder.* Like Christo's *Running Fence,* the video will be seen by many more people than ever experienced the actual simulation. If VR is a simulation of real life, what is a video production that simulates VR? Is it not a simulacrum, a copy without an original? The aura of originality that was traditionally important, as Walter Benjamin has observed, in establishing the authenticity of an artwork is here almost completely dissipated, for the video is as much or

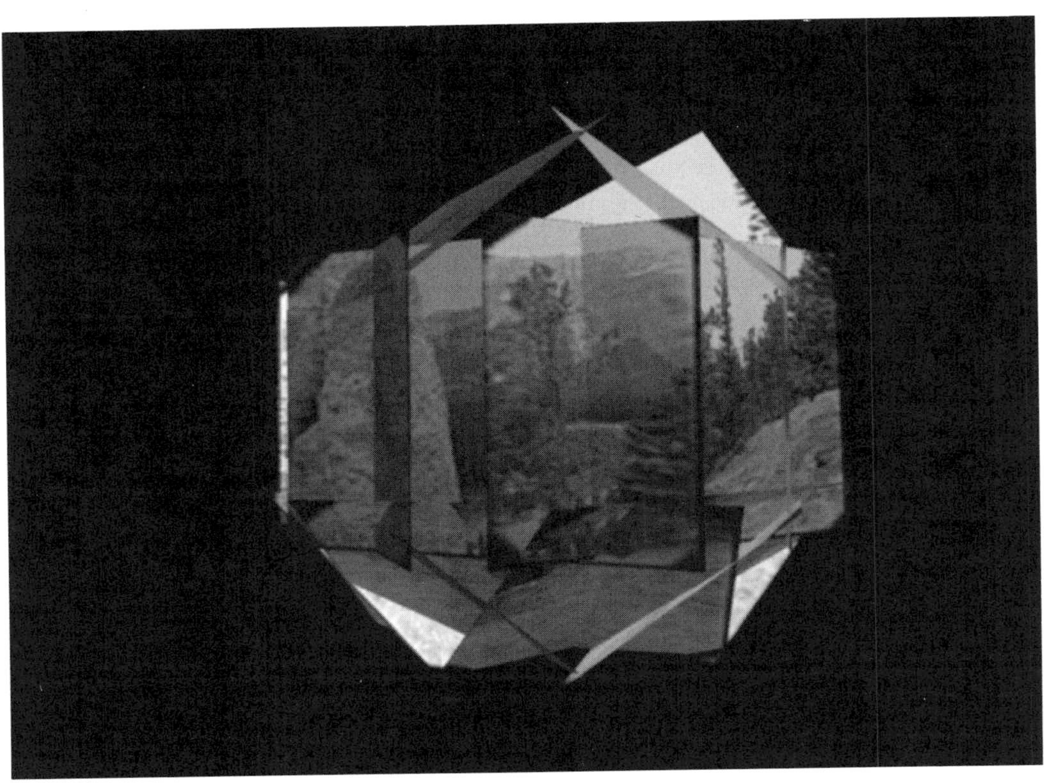

7. Brenda Laurel and Rachel Strickland, *Placeholder* (1993), video image of the composition of a three-dimensional environment. Digital image courtesy the artists.

more the work of art as the VR production itself. If there is a lesson to learn in this precession of simulacra, surely it is that once materiality and information are intertwined, neither is likely to manifest itself in any simple or unambiguous way. The simulation emphasizes embodiment precisely through its artificiality, and the video captures immediacy through its mediation. When matter and information begin to copulate, more is destabilized than contemporary constructions of the body. Also at issue are traditional categories of art criticism, including orginal/copy, artist/technician, and copyright holder/paid labor. Just as it is no longer sufficient to think of the body as flesh or code alone, so it is no longer sufficient to accept distinctions that rely on putting artists in one category and technicians in another, or originals in a gallery and copies in Walmart. It is not only the human body that is undergoing a sea-change. Also in the throes of mutation is the body of art.

In the video, also named *Placeholder,* one clip shows Laurel explaining to her colleagues that if the simulation is not up and running by the time that executives from Interval Research (her employers) show up at Banff, she will be out of a job. The scene poignantly captures the stress of trying to create an ambitious artwork within the constraints of a for-profit company. Unfinished, imperfectly documented, commercially unviable, and captive to capitalist financing, *Placeholder* testifies to the importance of embodied action in many senses. In a virtual world, it is as real as it gets.

Whereas Laurel and Strickland emphasize the materiality that interpenetrates virtuality, the Canadian artist Catherine Richards focuses on the virtuality that interpenetrates materiality. Her work explores the extent to which virtuality has overtaken RL—the real life that VR researchers shorten to an acronym, semiotically declaring its equivalence to VR. In her art video *Spectral Bodies,* Richards focuses on proprioception, the sensory system that uses internal nerve receptors at joints and muscles to give us the sense that we inhabit our bodies.[21] Normally we know, without thinking about it, exactly where our body boundaries are and how to move our bodies to negotiate complex spaces and topologies. Precisely because it is automatic, the importance of proprioception normally tends to be veiled from us. In the video Richards alludes to Oliver Sacks's essay

"The Disembodied Woman," in which he writes about Christine, a patient who lost her proprioceptive sense as a result of neurological damage.[22] With great effort, Christine was able to relearn how to sit, walk, and stand; what she was not able to recapture was the sense that she was inside her body. She felt that she was positioned somewhere outside, forced to manipulate her body through conscious effort as if she were a puppetmaster moving an inert doll.

Sacks's essay is only one of many sources that *Spectral Bodies* draws upon in exploring the possibility that proprioception, far from being secure except in instances of rare neurological damage, can quite easily be manipulated to give radically different experiences of body boundaries. The video includes footage shot in the laboratory of Mark Green at the University of Alberta. In the laboratory, Catherine Richards conducts a demonstration showing that the perceived boundaries of the body can be remapped by the low-tech method of stimulating the surface of the arms and hands with a vibrator. Already interested in VR through her pioneering work in the Banff Centre's Bioapparatus project, she was quick to see the connection between virtual experience and this proprioceptive remapping of the body surface. The video shows her enacting the experiment with various blindfolded subjects, who describe what they feel happening to their bodies. One woman reports that she feels her neck shrinking and then thickening, becoming like a bull's neck. The process continues until her head retreats into her chest, her shoulders forming an unbroken line across the top of her body. Another woman remarks that her arms are growing longer and longer, stretching six feet or more away from her body.

Combined with these images are visual and verbal references to virtual reality. As the video begins, lettering on the screen informs us that Richards is combining "body illusions with VR to bring to the surface the intervention of VR in the psychological (re)mapping of our bodies." When I spoke with Richards about the video, she pointed out that all of film, television, and video evolved from a single "perceptual glitch"—the fact that when the human eye is presented with images in rapid succession, it perceives the images as motion. She sees a similar plethora of technologies evolving from the "perceptual glitch" she explores in her video—the fact that our sense of body boundaries can easily be destabilized and reconfigured by even low-tech interventions. When the intervention consists of stimulating surface muscles while the subject is blindfolded, the reconfiguration is easily detected as a hallucinatory experience. The subjects in the video knew that their experiences were not "real," because ordinary experience as well as other sensory channels told them that the body was in fact remaining stable. Imagine, however, a VR simulation that would visually reinforce the changing image the subject has of her body. What if the woman not only felt proprioceptively that her neck was shrinking, but also saw it shrinking through VR eyephones? What if she interacted with her environment through this changed form, for example by trying to wear a hat and having it fall off because she had no exterior head? Then the borderline between hallucination and reality would become harder to detect. It is not difficult to imagine a technological intervention powerful, sustained, and consistent enough to remap a subject's body with compelling intensity.

Spectral Bodies is not an argument or an exposition, although it contains elements of both. Rather, it is an intervention that comes to the viewer through multiple channels, combining print, images, sounds, and narrative in a suggestive collage that hints at a synergistic collaboration between the informational and material body. In my conversation with Richards, she remarked that she did not think the interpenetration of materiality by information was primarily due to the development of VR technology. Rather, she sees VR technology developing because we who inhabit first world countries are already convinced of our virtuality. Everything else is becoming virtual—money is displaced by automatic teller machine (ATM) cards, physical contact by phone sex, face-to-face communication by answering machines, an industrial economic base by information systems; why shouldn't our bodies be virtual too? Yet even as she foregrounds this shift and wants to explore it, Richards insists on the continuing importance of our material existence. It is for this reason that she locates her work at the site of the body, because she sees the body as the crossroads where the psychological, material, and virtual intersect. The feminist edge to Richards's work comes not only from her insistence on materiality but also from her implicit assumption

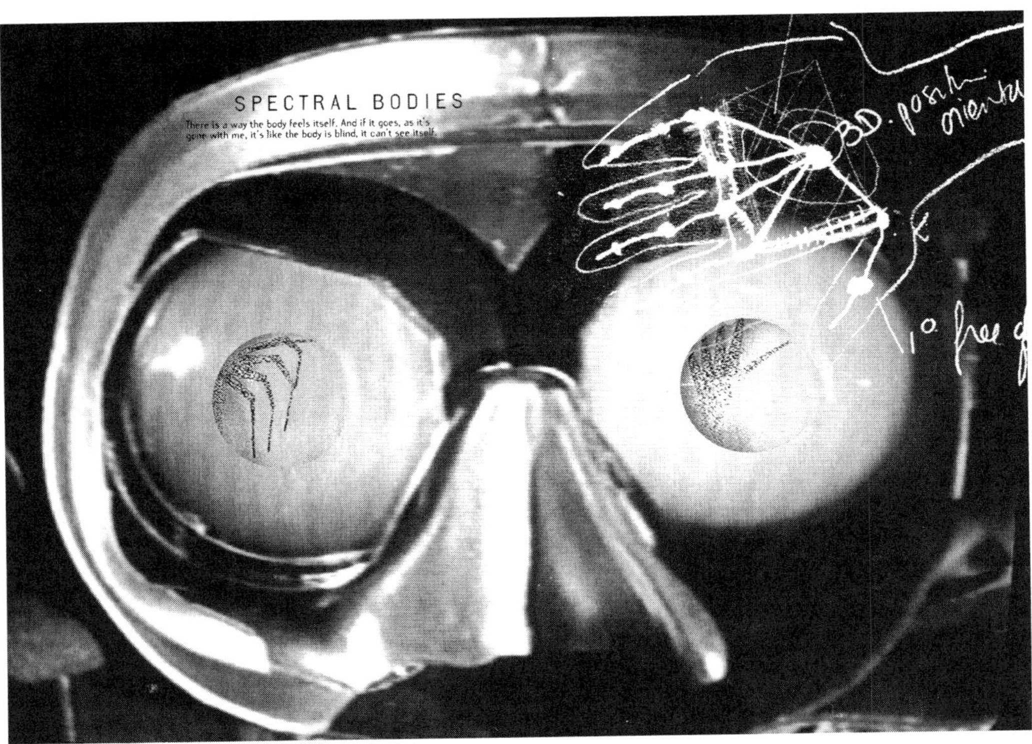

8. Catherine Richards's *Spectral Bodies* videotape, created in 1991, includes a segment of virtual reality in a set of stories about the loss of bodily presence. By embedding this component of virtual reality among other narratives, the artist treats virtual reality as one of a number of ways in which body boundaries can be destabilized.

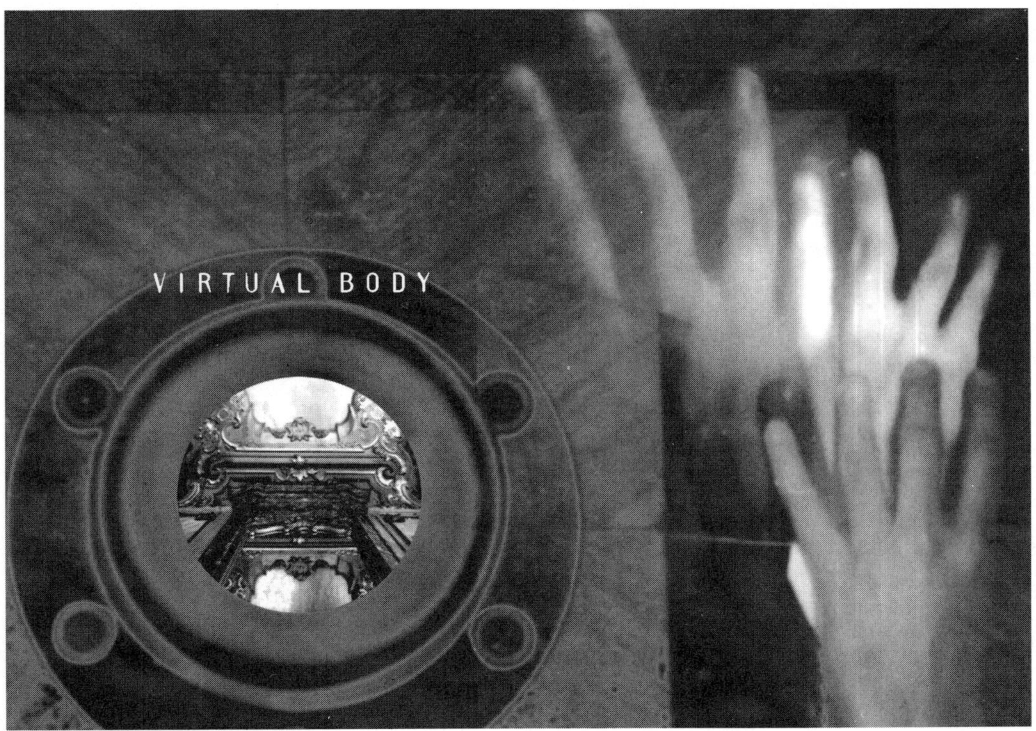

9. *The Virtual Body*, created in 1993, encapsulates artist Catherine Richards's concerns with the relationship between the body and new technologies. What appears to be, at first glance, a nostalgic material object is, in fact, an interactive artwork that blurs the boundaries between spectator and object.

that body and mind interact and that any reconfiguration of the body must necessarily affect how subjectivity is constituted. In this sense she has more in common with Laurel than she does with Moravec or Gibson.

Spectral Bodies ends with the voiceover of a subject who experienced radical body distortions asking, "Can you put me back [the way I was]?" As the VR artists discussed here realize, it is not possible for us in first world countries to return to a state of "natural" wholeness. I believe Laurel, Strickland, and Richards are correct in seeking to find a way *through* these powerful technologies rather than condemning or boycotting them. "I would rather be a cyborg than a goddess," Donna Haraway proclaims at the end of "A Manifesto for Cyborgs."[23] The question for us now is, what kind of cyborg? Dreams of transcendence, of freedom from the flesh, exacerbate rather than solve our problems. Although our bodies may not be "natural," they are still material, and they are still necessary for our survival on this planet. However resurfaced with virtuality, the physical world is our one and only home.

Notes

1. Hans Moravec, *Mind Children: The Future of Robot and Human Intelligence* (Cambridge: Harvard University Press, 1988).

2. The point is eloquently made by Francisco J. Varela, Evan Thompson, and Eleanor Rosch in *The Embodied Mind: Cognitive Science and Human Experience* (Cambridge: MIT Press, 1991).

3. For a discussion of these gender identifications, see Carolyn Merchant, *The Death of Nature: Women, Ecology and the Scientific Revolution* (San Francisco: Harper and Row, 1980).

4. Nancy Leys Stepan, "Race and Gender: The Role of Analogy in Science," *Isis* 77 (1986), 261–277.

5. Erwin Schrödinger, *What Is Life? with Mind and Matter and Autobiographical Sketches* (Cambridge: Cambridge University Press, 1944; reprinted 1992).

6. Gregory Bateson, *Steps to an Ecology of Mind* (reprinted New York: Ballantine Books, 1972).

7. Norbert Wiener, *Cybernetics: Or Control and Communication in the Animal and the Machine* (Cambridge: MIT Press, 1948, 2d ed. 1961), l.

8. William Gibson, *Neuromancer* (New York: Ace Books, 1984).

9. The phrase "embodied virtuality" was coined Mark Weiser of Xerox PARC. I am indebted to Mark Weiser for making it possible for me to visit Xerox PARC and see these technologies firsthand.

10. A. J. Greimas, "The Interaction of Semiotic Constraints," *On Meaning: Selected Writings in Semiotic Theory,* trans. Paul J. Perron and Frank H. Collins (Minneapolis: University of Minnesota Press, 1987), 48–62. See also Ronald Schleifer, *A. J. Greimas and the Nature of Meaning: Linguistics, Semiotics and Discourse Theory* (London: Croom Helm, 1987), 22–55. Also helpful is the discussion of the semiotic square in Ronald Schleifer, Robert Con Davis, and Nancy Mergler, *Culture and Cognition: The Boundaries of Literary and Scientific Inquiry* (Ithaca: Cornell University Press, 1992).

11. The classic text here is Jacques Derrida, *Of Grammatology,* trans. Gayatri Chakravorty Spivak (Baltimore: Johns Hopkins University Press, 1974).

12. The explanation of the relation of information to pattern/randomness, condensed here for reasons of space, can be found in fuller form in N. Katherine Hayles, "Virtual Bodies and Flickering Signifiers," *October* 66 (Fall 1993), 69–91.

13. Claude E. Shannon and Warren Weaver, *The Mathematical Theory of Information* (Urbana: University of Illinois Press, 1947).

14. Jean Baudrillard, *Simulations* (New York: Semiotext(e), 1983). See also Jean Baudrillard, *The Ecstacy of Communication* (New York: Semiotext(e), 1988).

15. For examples of the emphasis on freaks and mutation in contemporary literature, see Katherine Dunn's *Geek Love* (New York: Knopf, 1989) and Tom DeHaven's *Freaks'*

Amour (New York: Penguin, 1986; first printing, William Morrow and Company, 1979). This aspect of contemporary literature is discussed in N. Katherine Hayles, "Postmodern Parataxis: Embodied Texts, Weightless Information," *American Literary History* 2 (1990), 394–421.

16. Shoshana Felman, *The Literary Speech Act: Don Juan with J. L. Austin, or Seduction in Two Languages,* trans. Catherine Porter (Ithaca: Cornell University Press, 1983), 141–142. See also Fredric Jameson, "Foreword," in Greimas, *On Meaning,* xvi.

17. Richard Dawkins, *The Selfish Gene* (Oxford: Oxford University Press, 1976).

18. The era of the posthuman has been discussed by Howard Rheingold in *Virtual Reality* (New York: Simon and Schuster, 1991) and Bruce Mazlish in *The Fourth Discontinuity: The Co-Evolution of Humans and Machines* (New Haven: Yale University Press, 1993).

19. These effects were reported to me in informal conversation with Stone and Laurel and anecdotally confirmed by researchers at the University of Washington Human Interaction Laboratory.

20. *Placeholder* is among the works presented in this book; Rachel Strickland and Brenda Laurel also showed the video on *Placeholder* at the Banff Centre for the Arts Symposium on Art and Virtual Environments in May 1994. I am indebted to the Banff Centre for making it possible for me to attend the conference and for their generous support and encouragement.

21. Quotations from *Spectral Bodies* are used with permission of Catherine Richards.

22. Oliver Sacks, "The Disembodied Woman," in *The Man Who Mistook His Wife for a Hat and Other Clinical Tales* (New York: Summit Books, 1985).

23. Donna Haraway, "A Manifesto for Cyborgs: Science, Technology, and Socialist Feminism in the 1980s," *Socialist Review* 80 (1985), 65–107.

Virtual Skin: Articulating Race in Cyberspace

Cameron Bailey

Is "race"[1] corporeal? Is that all there is to one of the most complex and contested discourses of the modern era—skin, eyes, lips, and hair? Clearly not. Most theories of race reject a biological basis altogether in favor of a tangle of social, political, and psychic forces that work their strange and funky work on each one of us every day. That is how it goes in the real world.

But what about cyberspace?[2] Do the same laws apply? Recent writings on electronic communication systems insist that despite its disembodied nature, cyberspace remains what Michael Benedikt calls a familiar social construct "with the ballast of materiality cast away."[3] This suggests that race may function there in much the same way as it does in the world where we are more directly accountable to our bodies. It may in fact mean this, but it is hard to tell, because very few of the thinkers currently probing into cyberspace have said a word about race.

Faced with the delirious prospect of leaving our bodies behind for the cool swoon of digital communication, many leading theorists of cyberspace have addressed the philosophical implications of a new technology by retreating to old ground. In a landscape of contemporary cultural criticism where the discourses of race, gender, class, and sexuality have often led to great leaps in understanding—where, in fact, they have been so thoroughly used as to become to some a mantra—these interpretive tools have come curiously late to the debate around cyberspace. It may be that the prevailing discussion of digitally assisted subjectivity has focused not on the culture of cyberspace as it exists today but on the potential of cyberspace, on utopian or dystopian visions for tomorrow. Since we never reveal ourselves so much as when we dream, it is worth noting that most speculations on the future of cyberspace return questions of race in particular to the margins. Volumes such as Michael Benedikt's *Cyberspace: First Steps* and Scott Bukatman's *Terminal Identity* barely mention the subject at all; only the work of writers like Donna Haraway and Vivian Sobchack has taken the question of cybernetic identity beyond a direct relationship between technology and a unified, representative, obvious human subjectivity.

But does race matter? Can it sustain itself in the shifting space of virtual communities? It would seem clear that the safety of binary oppositions—self/other, black/white, male/female, straight/gay, writer/reader—would evaporate in the forcefully uncertain world of electronic discourse. A message comes and goes without a face, communication takes place without bodies to ground it, to provide the deeper layers of meaning below the surface upon which we all depend. This is especially important given the extent to which social interaction depends on embodied communication, on stable, known genders, sexualities, races, and classes somewhere present in the communicative act. Without this there would be no power flowing through communication, and without the flow of power, what would we have to say to one another?

Cyberspace communication challenges all that. In the online world, identity is often chosen, played with, subverted, or foregrounded as a construct. There appears to be a demonstration of the freedom provided by disembodied communication, the ludic element that is central to cyberspace activity in general,[4] as well as the influence of twenty-five years of postmodernity. What makes cyberspace so interesting as a public sphere is how none of the usual landmarks can be trusted. Also, the old economy of readers and writers, speakers and listeners is turned sideways; with the simultaneity and multidirectionality of online communication, authority is won and lost with such frequency that it becomes nearly irrelevant.

However, online interaction is anything but a utopia of democratic communication. Feminist critics such as Allucquère Rosanne Stone, Sally Prior, and Jill Scott have pointed out how cyberspace is gendered to reproduce boring phallocratic limits on expression. Many have noted that the ideal of unfettered democracy touted by so many champions of the Internet contains its own ideological dead weight. Net advocates often seem trapped by the boundaries of Enlightenment notions of individual freedom on one hand and Marshall McLuhan's utopia of communication on the other. Like the democracy of the ancient Greeks, today's digital democracy is reserved for an elite with the means to enjoy it. So it is with race. Existing racial discourses find their way into cyberspace, not simply as content but as part of the structure shaping the place. As with any other arena where identities are produced and exchanged, this aspect of cyberspace rests on the question of representation.

I want to look at issues of representation at both the social and personal level, to distinguish between what Kobena Mercer and Isaac Julien call representation as delegation and representation as depiction. In social terms, it is necessary to examine how variant communities are constructed online and what kind of access different communities have to communication technology. In the United States, for instance, there is a growing movement among African Americans to resist exclusion by corporations getting ready to wire the suburbs for the forthcoming ideology—a.k.a. information—superhighway. While this is primarily a consumer issue that only grazes deeper questions of engagement with the apparatus, there comes with this mobilization a push for greater technological literacy among blacks and other disenfranchised people.

In personal terms, we need to explore what it means to construct identity without the aid of racial and cultural markers like physical appearance, accent, and so on. Here I will be dealing exclusively with those forms of electronic communication that depend on text instead of any figurative representation of the physical body—that is, Internet newsgroups, online forums, e-mail, and text-based environments like Multi-User Dungeons (MUDs). On the surface it would seem that these are literary domains that function like an exchange of correspondence or the letters page of a newspaper.[5] One *presents oneself* in language, as is done in all forms of writing, which require the multifaceted acts of identity construction, selective editing, and telling of lies. But online communication adds something more: speed and uncertainty. MUDs operate in close-to-real time, providing an instantaneity that remains disembodied like writing but is nonetheless immediate like the telephone. And the literary contract between writer and reader becomes blurred. In the world of Internet newsgroups, mailing lists, and electronic bulletin board systems (BBSs), writers post messages simultaneously to individuals and to groups sharing a similar interest. The question of address becomes more complex. Also, the way in which these messages are retrieved and read gives the reader a power akin to the hiphop sampler's authority over source music—it is a consumer's market. All of this uproots the online writer's sense of his or her centered self. If identity is created solely through text and the text is as fluid as this, things fall apart in interesting ways.

My entry points for exploring the special glow of virtual skin are shaped, first, by the perspective of an online browser who has been involved in local BBSs, like Matrix and Magic in Toronto, the CompuServe commercial network, as well as the less-regulated Internet; and, second, by a continuing interest in the formation of new communities. Like all good postmodern citizens, I have learned to move with shifts in imagined communities, to ride the knowledge that, as Allucquère Rosanne Stone notes, "technology and culture constitute each other."[6] I may not swim, but I have learned to surf.

My first experience of virtual community came in Rock Dundo, Barbados, 1969,[7] when I first jacked in to a smooth, plastic, khaki-colored ViewMaster™. My mother, thousands of kilometers away in Canada, sent me both the machine and its software—disks that brought to life before my eyes images I had never seen before: Niagara Falls and Flowerpot Island and Toronto City Hall in stereoscopic vision. It would be two decades before I tried on a helmet, but I knew the thrill of virtual reality right then. I was *transported*. Every time I returned to that machine I left the postcolonial sunshine behind for the marvels of Canada. Immersed in the depth, resolution, and brightness of those images, I became a part of Canada, sharing an experience with every tourist who had paused to get a good look at new city hall, who had marveled at the falls. More importantly, by entering these images I could share the desire for the spectacle of Canada with my mother, who had recently immigrated there.

Now, producing these words on a newer piece of fetish hardware—a matte-black IBM ThinkPad™—I can extend into corners of cyberspace, remaking myself by will and accident, reading and misreading others. It is exhilarating at first, but it is not new. As Stuart Hall and others have pointed out,[8] migration is a central part of the postcolonial experience, and it necessarily involves shifting identity. It is the nature of Asian and African new-worlders to pass through different allegiances, belief systems and accents—for me it was Wembley, Rock Dundo, and now suburban Toronto—as a common part of life. At the same time, one develops a hyperawareness of the relationship between physicality and identity. Like women, like lesbians and gays, people of color[9] living in western metropoles live a crucial part of their existence as body-people, as subjects named and identified through their flesh. One need only hear "Monkey!" or "Water buffalo!" screamed at you on the street every once in a while to be reminded of that.

Esprit de Corps

> I am not that set of limbs called the human body.[10]
> René Descartes

Descartes has caught unimagined hell from countless thinkers for dividing the self into mind and body functions so cleanly. Cybertheorists return to his work not only for its mapping of space, but also for its notion of the split subject, consciousness split from flesh. In selected bits, Descartes does indeed appear binary:

> By *body* I mean whatever is capable of being bounded by some shape, and comprehended by some place, and of occupying space in such a way that all other bodies are excluded; moreover of being perceived by touch, sight, hearing, taste, or smell; and further, of being moved in various ways, not of itself but by some other body

> that touches it. . . . The power of self-movement, and the further powers of sensation and consciousness (*sentiendi, vel cogitandi*) I judged not to belong in any way to the essence of the body.[11]

However, in the process of his argument one finds a struggle to name the differences between mind and body, a struggle that belies his more definitive conclusions. What remains important is that indeterminacy. Consciousness meets corporeality in countless guises; sometimes they swap clothes.

One of the favorite playgrounds for mind/body has always been art. The experience of visual, literary, and musical forms allows a projection of consciousness into something else, an out-of-body experience that becomes especially heightened with figurative and narrative forms. When we play a video game we in part become a self beyond ourselves—jet fighter, Street Fighter, or Sonic. When we watch a film, particularly a narrative film, we identify with the protagonists, we stitch ourselves into the narrative. The persistence-of-vision known as Harrison Ford becomes me. A similar process occurs with written fiction. In each case a combination of imagination and image-producing technology works to project a self outside our bodies.

Cyberspace projection differs in important ways. Online, the subject is more fully responsible for the persona into which she or he projects. That virtual self no longer comes ready-made from Hollywood, or Harlequin, or Nintendo. It is a product of one's own words and acts. Also, the metaphor of cyber*space* emphasizes the immersive quality of the experience: even more than with reading or with cinema spectatorship, one *gets inside*. While cinema permits an illusion of participation, cyberspace is predicated on it. From the very first computer games, the graphical Spacewar and the text-based Adventure, the model of the disembodied, simulated subject moving through Cartesian space, what Scott Bukatman calls the cybersubject, was already in place.[12]

But there is an illusion at work in the immersion model of cyberspace and virtual reality. As Sally Pryor and Jill Scott note, despite the participatory nature of VR, this spatial metaphor still emphasizes "the Self behind a 'camera' looking out a window on the world."[13] The cybersubject remains distinct and apart from his or her virtual environment, which is a result both of the nature of Cartesian space, which requires a vantage point outside of the spatial field, and of the primacy, at least in virtual reality, given to looking *at* and moving *forward*. So here we are, smack up against ideology: the metaphors developed to describe the experience of cyberspace are made from the same western, masculinist ideals previously dismantled by this quarter-century of poststructuralist thought. The cyberspatial ideal of freeing, disembodied, decentered communication is effectively countered by an imaginative system that reinscribes a neat binary opposition between self and other.

The cybersubject as defined by most current theorizing is not only gendered but also has a clear cultural specificity that derives from a calcification of the questions that run through Cartesian thought. Steven Whittaker defines the typical cyberspace enthusiast as someone who "desires embodiment and disembodiment in the same instant. His ideal machine would address itself to his senses, yet free him from his body. His is a vision which loves sensorial possibility while hating bodily limits. He loves his senses and hates his body!"[14] It is as lurid as "I Was a Teenage Cyborg," though not so innocent. Pryor and Scott remind us of the link between this mind/body split and related oppositions like self/other, subject/object, and male/female; they also insist on remembering the value-laden nature of these oppositions, one side desirable, the other dead weight: "It is not surprising that the body, subject to vulnerability, pain and mortality, can become something from which it seems desirable to escape. Could you feel pain if you had no body? Could you experience racism or sexism?"[15]

So, taking the preferred side of the handful of primary couplets of identification, the cybersubject as currently figured is male, white, straight, able-bodied, and ruling-class. So what? Any identity that occupies the shadow half of these categories (female, black, queer) remains lashed to his or her body. Libraries of feminist thought tell us that a woman's identity has historically been defined and maintained through the body. The same holds true for Africans in the

West, aboriginal people, and so on. Biology is destiny. Physiology is law. Subjecthood lies over the horizon. This becomes especially interesting in a domain that privileges the forfeit of the body so eagerly. That process is neither universally simple nor universally desirable.

It is important to distinguish here between the cybersubject as a figure produced by current thought about cyberspace and the actual people who enter cyberspace every day. In the same way that film theory distinguishes between the cinematic spectator as a function of the cinematic text and "real world" viewers of movies, we must note that the cybersubject defined above is produced by still limited notions of the experience of cyberspace, and has a relationship to, but is in no way coextensive with, the millions who communicate online or enter virtual reality. Cyberspace is built for that unified subject, but inhabited by a happily chaotic range of subjectivities.

Freeing up movement, communication, and sensation from the limitations of the flesh might be the promise of digital experience, but the body will not be abandoned so easily. Western culture concurs that the quality of imagination is what allows all manner of disembodied experience, from being "immersed" in narrative to the spatial metaphors of cyberspace. Returning to Descartes, his notion of imagination appears suddenly pertinent: an "application of the cognitive faculty to a body intimately present to it—a body, therefore, that exists."[16] An awareness of the physical, "real" body is crucial to the disembodied projections of cyberspace. The physical body remains as a referent. Cyberspace wouldn't make sense without it. Here lies the connection between race and cyberspace. Western racial discourse began in a scientistic attempt to account for physical differences among people.[17] Even when its meaning had left any pretense at science behind and extended into social and political spheres, the fact of the body remained. Skin, eyes, lips, and hair endured as a powerful referent, ready to be drawn upon as evidence. At its most abstract, racial discourse still involves an imaginative act that relies on the physical body. Habeas corpus, or there is nothing to discuss.

SHAREWARE

> Umntu ngumntu ngabantu—a human being is a person through (other) people.[18]
> Bantu proverb

The cyberspace nation is in the house. With its mail, discussion groups, bulletin boards, and shareware, with its geography and its idiom, cyberspace simulates community, a community more dependent on imagination than most. Benedict Anderson maintains that "all communities larger than primordial villages of face-to-face contact (and perhaps even these) are imagined. Communities are to be distinguished, not by their falsity/genuineness, but by the style in which they are imagined."[19] In his schema, a nation coheres around three principles: (1) to be limited: "No nation imagines itself coterminous with mankind";[20] (2) to be sovereign; and (3) to be a community: "Regardless of the actual inequality and exploitation that may prevail in each, the nation is always conceived as a deep, horizontal comradeship."[21]

So what is the nature of the online community? First, the economics of online communication require that participants have access to a computer, a modem, and a telephone line. Cancel tens of millions of North Americans. Until recently, Internet access required membership in an elite institution—a university, government department, or major corporation. Millions more gone, but not evenly across the board. In the United States, African Americans and Hispanics are overrepresented among those without Net access, as are aboriginal people in Canada. Owning the means of participation is a class issue, and another example of how class is racialized in North America. In writing about poverty and information, Karen G. Schneider argues that "the information-rich, however well-meaning, have largely determined and prioritized the issues of the information revolution according to their own visions and realities."[22] What happens when the class of the information-rich is also racialized, when it continues to be predominantly white?

Beyond economics, there is a somewhat harder-to-quantify culture of cyberspace. The Net nation deploys shared knowledge and language to unite against outsiders: Net jargon extends beyond technical language to acronyms both benign (BTW, "By the way") and snippy (RTFM, "Read the fucking manual"). It includes neologisms, text-graphical hybrids called emoticons, and a thoroughgoing anti-"newbie" snobbery. Like any other community, it uses language to erect barriers to membership.[23] As Anderson also suggests, print culture is crucial to the formation of nations. The Internet is nothing if not a riot of publishing, often about itself. Popular guides like Brendan Kehoe's *Zen and the Art of the Internet,* as well as the countless lists of Frequently Asked Questions (FAQs), serve to provide a body of common knowledge and therefore enforce order on the Net. There is in these codes of language, and in the very concept of "netiquette," something of the culture of suburban America; one gets the sense that these structures are in place not simply to order cyberspace but to keep chaos (the urban sphere) out. It is no stretch to suggest that in turning to cyberspace, the white middle-class men who first populated it sought refuge from the hostile forces in physical, urban space—crime, poor people, desperate neighborhoods, and the black and brown. In writing about a BBS called New York Online, Noah Green compares the hermetic concerns of most BBSs to white flight from urban reality. NYO promises "a virtual community that's a complement to, not an escape from, an existing physical one."[24] Departing from the usual thrill of online communication—erasing huge global distances through instant, intimate connection—NYO emphasizes an electronic closeness that derives from a geographical one. Most of its membership—"50 per cent minority and 40 per cent female"—lives in Brooklyn. The norm, however, remains closer to Michael Heim's vision of the alienated subject-under-siege:

> Isolation persists as a major problem of contemporary urban society—I mean spiritual isolation, the kind that plagues individuals even on crowded city streets. . . . For many, networks and bulletin boards act as computer antidotes to the atomism of society.

> Unfortunately, what technology gives with one hand, it often takes away with the other. Technology increasingly eliminates direct human interdependence. . . . Because machines provide us with the power to flit about the universe, our communities grow more fragile, airy, and ephemeral, even as our connections multiply.[25]

So the suburban ideal of postwar North America returns in virtual form: communication at a safe distance, community without contact. Is it any wonder that when movies visualize the Net's matrix of communication,[26] it so often resembles the cool, aerial patterns of a suburb at night?

One often overlooked contributor to Net culture is the ludic aggression of adolescent masculinity. We have seen how cyberspace is gendered as masculine, but the community of hackers, late-night Net surfers, BBS sysops, and virus writers has often included large numbers of teenagers. Particularly since the era when popular culture first came to be identified with teen culture, adolescence and especially male adolescence has been accorded profound importance and created a profound disturbance in western society—just look at all the mechanisms in place to control it. In acts both constructive and transgressive, adolescent boys have used cyberspace to express the flux, despair, anger, restlessness, and pain of coming to adulthood. In doing so, they have shaped the character of online community to reflect secrecy, game structures, and hostility to authority. The sense of combative play engendered by this group extends the range and focus of the imaginative act that entry into cyberspace requires. Stone points out that

> many of the engineers currently debating the form and nature of cyberspace are the young turks of computer engineering, men in their late teens and twenties, and they are preoccupied with the things with which postpubescent men have always been preoccupied. This rather steamy group will generate the codes and descriptors by which bodies in cyberspace are represented.[27]

In terms of racial discourse, an interesting relationship is established between young white men and the sizable numbers of Asian American and Asian Canadian teenage boys who have also contributed to the development of Net culture. The closest parallel is with indie rock and 'zine culture, which are also populated by a predominantly, but neither exclusively nor agressively, white teen tribe. In both cases, Asian youth participate according to the terms of the subculture, which demands a cultural "neutrality." Black youth, with their own clearly defined and visible youth culture, must engage in a more complex negotiation.

An earlier generation of North American counterculture has also left its mark on cyberspace. Philip Hayward identifies the technology's leading advocates as "a specific social group comprising individuals who have clung to residual 'counter-cultural' notions, most often articulated within terms of a loose Green-Libertarian rhetoric, while being assimilated into certain sectors of the American professional classes."[28] John James's pioneering BBS CommuniTree fit at least part of that description: it started up in the late 1970s as a virtual community[29] with principles and language drawn from the Aquarian age. Ironically, it was infiltrated and eventually destroyed by adolescent hacker boys.[30]

Following from Benedict Anderson, we can say that the online nation has constructed itself as a community that is not racist by stated principles but, because of the way nations are always constructed, has built affinities (and, by definition, exclusions) that have the effect of shunting aside certain voices, languages, and vernaculars. However, this historical condition is now in tremendous flux as the online world grows to become a collection of communities. *Time* magazine has shrieked that "now that the population of the Net is larger than that of most countries in the world . . . the Internet is becoming Balkanized."[31] I prefer to see the change as more in keeping with the established, decentralizing spirit of the Net. Now at a transitional stage before commerce stomps in, cyberspace is more open to the free play of subcultures than it ever was. Some examples:

- Soc.culture.african.american is one of the busiest of Usenet newsgroups, accumulating hundreds of posts every few hours.

- Dozens of other newsgroups are devoted to a variety of self-defined cultural communities. The speed, anonymity, and diffusion of newsgroup debate mean that subjects usually confined to safe, private conversation among friends or family are given semipublic airing on Usenet. Genocide theories and interracial dating are perennials in soc.culture.african.american; everything from assimilation to eating dogs comes up in soc.culture.asian.american.
- In addition to this kind of debate, aboriginal activists use alt.native and soc.culture.native to get the word out on local struggles and call for support from the online community.
- African American cyberspace activist Art McGee compiles and distributes regular surveys of mailing lists, newsgroups, and BBSs of interest to African Americans. The catalogue of mailing lists numbers more than sixty, including lists devoted to the Association of Black Sociologists, Cameroonian students studying in London, and departed jazz guru Sun Ra. McGee's signature line is "The revolution will not be televised, but the proceedings will be available online."
- NativeNet, an online network organized in part by aboriginal artists working through the Banff Centre for the Art, spans North America.
- Dozens of black-specific electronic bulletin boards have sprung up across North America, including: Black Board International in Toronto, Ontario; Imhotep in Brooklyn, New York; Pan-Africa Online in Pasadena, California; and Girlfriend! in Arlington, Virginia. Many of these BBSs are linked through a network called Afronet.[32] Afronet has recently been joined by Melanet in linking people of African descent in cyberspace.
- The Russell County BBS was created in Hobson, Montana, designed as a meeting place and native art gallery. Russell County is one of a small number of bulletin boards using NAPLPS (North American Presentation Level Protocol Syntax) to compress and distribute First Nations visual art and children's animation.[33]
- The sale and exchange of digitized porn images caters increasingly to racial fetishes, with white and Asian women pictured in interracial

scenarios carrying the highest currency. The narratives of interracial desire remain popular on porn BBSs, and even on African American porn BBSs like Ebony Shack, images of black male/white female scenarios sometimes outnumber all other configurations.

- As aboriginal people and people of color organize online, so do far-right organizations. According to Reuters and *U.S. News and World Report,* neo-Nazi hate literature has been discovered by browsers on bulletin board systems in Germany, Sweden, France, and the Netherlands.

The social dynamics of Usenet culture in particular encourage subcultures; with its devotion to trading arcane knowledge and to the same celebration of spontaneous opinion that one finds all over North American talk radio, this medium is tailor-made for generating communities within communities.

Digitalia

> I occupied space. I moved toward the other . . . and the evanescent other, hostile, but not opaque, transparent, not there, disappeared. Nausea.[34]
>
> Frantz Fanon, *Black Skin, White Masks*

The discourse of race is, by history and by design, rooted in the body. Cybersubjectivity promises the fantasy of disembodied communication, but it remains firmly connected to bodies through the imaginative act required to project into cyberspace. What cybersubjectivity actually offers is reembodied communication. So how should I reembody myself amidst the Net's possibilities for self-presentation? Where should I look for my digitalia, that odd conflux of intimacy (genitalia), foreignness (marginalia), and wires? Should I announce myself racially, give myself a secure racial identity? As an experiment, I conducted a poll in CompuServe's African American forum, asking how participants situated themselves online:

> More often than not I do not identify myself when I interact with people except in forums such as this one. Why should I, really? I have had more negative experiences with people being overtly racist in cyberspace than I have in FTF [face to face] life. I find it intriguing to experience what people will tell me when they think I am White.
>
> In the other CompuServe forums and Usenet newsgroups which I frequent, I encounter a lot more racist (and sexist, and homophobic, and anti-Semitic, and otherwise bigoted) messages than in "real life." I think the anonymity of on-line communications is very enticing to bigoted fools.
>
> Here's a thought: Do you think bigoted people are attracted to cyberspace, or are "normal" people encouraged to show their hidden bigoted sides?
>
> I have heard people making derogatory comments about Mexican Americans, Asians, Gays, Lesbians, and Bisexuals, etc . . . and although I am not a member of those groups, I feel it is essential that I confront intolerance, period. So I suppose letting people know who I am is not as important to me as letting people know what I will not put up with.

What was most interesting about the response was how quickly the thread moved away from the question of how one identifies oneself to a more manageable debate about racism. From what I have been able to glean in this and other online conversations, many African Americans (my survey was limited in sample) are unwilling to probe too deeply into the part racial identity plays in their conception of themselves, into the part of them that stays black when they present no "evidence" of blackness. Race is either taken for granted or deliberately left unspoken. In a GEnie conference on African American access to information technology, a quiet consensus emerged on the value of racial anonymity online:

> One nice thing about online communication is that everyone is equal; no one knows how old a participant is, or what color, or what gender, or what religion—which frees our minds a bit to listen to more diverse opinions.

Another participant commented:

> When you type away, no one online need know your skin color. Accents don't matter as much. High-tech is a wonderful way to fight snobbery!

Given that cyberspace is a racialized domain, this sort of virtual transvestism is by no means neutral. In an another era it used to be called passing. There is another option. Taking a cue from the adolescent boys who determined so much of cyberculture, I could play. I could try to extend my engagement with cyberspace beyond the ludic economies of North American teenagers to include trickster traditions, signifying, and elements of spirituality that lie outside western rationalism. That way subjectivity need not be a fixed racial assertion nor a calculated transvestism; it could be more fluid, more strategic. William Gibson was the first to write about variant cosmological approaches to cyberspace, contrasting his protagonist Case with the Rastafarian-derived Zionites in *Neuromancer,* and making extensive uses of vodun in *Count Zero* and *Mona Lisa Overdrive.*[35] While this offers enormous possibilities, there is a danger, at least in fiction, of surrendering to the same sort of essentialism that defines people of color in exoticizing, body-oriented terms. Michael Heim, for instance, in lamenting cyberspace's retreat from the physical body, offers Gibson's Zionites as a symbol of salvation:

> Gibson leaves us the image of a human group who *instinctively* keeps its distance from the computer matrix. These are the Zionites, the religiously *tribal folk* who prefer music to computers and intuitive loyalties to calculation. . . . As we suit up for the exciting future in cyberspace, we must not lose touch with the Zionites, the *body people* who remain *rooted in the energies of the earth.*[36]

In Gibson's novel, the Zionites are rooted in both technology and spirituality. But taken by Heim as a symbol, they are reduced to "body people."

I prefer to go all the way back to that View Master™, holding it up to the bright Barbados sun so I could see Canada better. Maybe this is an answer: the ecstasy of projected community and irresolvable difference, both claimed in the very same moment.

NOTES

The research and writing of this article was assisted by a grant from the Ontario Arts Council's Arts Writing Program.

1. As an initial flag, I place the word "race" in quotation marks to acknowledge the work that Henry Louis Gates, Tzvetan Todorov, and others have done to explore how race is a constructed discourse, not a biological or even a social fact. However, I do not believe that quotation marks resolve the question—what do we do with all of language's other slippery concepts?—so from now on, I will leave race to fend for itself.

2. My focus will be on the domain of online communication: bulletin board systems, commercial online services, and the Internet; that, is the aspect of cyberspace that exists as a public sphere.

3. Michael Benedikt, ed., *Cyberspace: First Steps* (Cambridge, Mass.: MIT Press, 1991), 4.

4. Scott Bukatman notes that virtual reality "is a ludic engagement with the space of the computer that refigures it as a perceivable and physical environment. Many of the designers of NASA's virtual reality system had, as a matter of fact, once worked as game designers for Atari" (*Terminal Identity: The Virtual Subject in Postmodern Science Fiction,* [Durham and London: Duke University Press, 1993], 200). That concept of play can be extended into how users navigate, sample information, and communicate via the Internet.

5. Allucquère Rosanne Stone makes important links between online interaction and more familiar forms of interactive, disembodied communication, particularly telephone conference calls and "communities of letters," in her essay, "Will the Real Body Please Stand Up?: Boundary Stories about Virtual Cultures," in Benedikt, *Cyberspace: First Steps,* 84.

6. Ibid., 82.

7. Coincidentally, this same year saw the start-up of CompuServe as the computer network of Golden United Investment, a life insurance holding company. In fact, the period from the late 1960s through the late 1970s, with the beginnings of the Internet and the development of the first computer bulletin boards, coincides with the final melting of first world national-ethnic borders, as more and more immigrants arrived in western metropoles from Asia, the Caribbean, Africa, and Latin America.

8. Stuart Hall, "New Ethnicities," *Black Film, British Cinema,* ICA Documents 7 (London: Institute of Contemporary Art, 1988), 27–30.

9. I will use "people of color" to refer to all those who identify themselves as being in whole or part of African or Asian descent. Though my focus will be on African American and African Canadian experience online, I want to include Asian and First Nations activity as important and necessary comparisons. The histories of Africans, Asians, and aboriginal people online are as different as they are off, but, as in the real world, the

experience of creating identities out of marginalization lends a common character to the process.

10. René Descartes, *Philosophical Writings: A Selection,* trans. and ed. Elizabeth Anscombe and Peter Thomas Geach (London: Open University, 1970), 69.

11. Ibid., 68.

12. Bukatman, *Terminal Identity,* 197.

13. Sally Pryor and Jill Scott, "Virtual Reality: Beyond Cartesian Space," in *Future Visions: New Technologies of the Screen,* ed. Philip Hayward and Tara Wollen (London: BFI Publishing, 1993), 168.

14. Steven Whittaker, "The Safe Abyss: What's Wrong with Virtual Reality?" *Border/Lines* 33 (1994), 45.

15. Pryor and Scott go on to speculate, only partly in jest, that cybertheory's fantasizing about escaping the body comes from male scientists completely out of touch with their own flesh, 172.

16. Descartes, *Philosophical Writings,* 109.

17. Lucius Outlaw, "Toward a Critical Theory of Race," in *Anatomy of Racism,* ed. David Theo Goldberg (Minneapolis: University of Minnesota Press, 1990), 61–68.

18. Joan Maw and John Picton, eds., *Concepts of the Body/Self in Africa* (Vienna: Afropub, 1992), 1.

19. Benedikt, *Cyberspace: First Steps,* 15.

20. Ibid., 16.

21. Ibid.

22. Karen G. Schneider, "Poverty and Information," Usenet post: kgs@panix.com.

23. Benedict Anderson, *Imagined Communities: Reflections on the Origin and Spread of Nationalism* (London: Verso, 1983), 133.

24. Noah Green, "The Sixth Borough: A Bulletin Board Grows in Brooklyn," *The Village Voice* (12 July 1994), 38.

25. Michael Heim, "The Erotic Ontology of Cyberspace," in Benedikt, *Cyberspace: First Steps,* 73–74.

26. Brett Leonard's *The Lawnmower Man,* 1992, and Rachel Talalay's *Ghost In The Machine,* 1993.

27. Stone, "Will the Real Body Please Stand Up?," 103–104.

28. Philip Hayward, "Situating Cyberspace: The Popularization of Virtual Reality," in Hayward and Wollen, *Future Visions,* 189.

29. Stone, "Will the Real Body Please Stand Up?," 88–92.

30. Ibid., 91.

31. Philip Elmer-Dewitt, "Battle for the Soul of the Internet," *Time* 144:4 (July 1994), 43.

32. According to Art McGee, "Afronet is an echomail backbone supported by African and African-American BBS sysops across North America. The goal is to distribute conferences with African and African-American themes throughout North America. It was originally conceived by Ken Onwere."

33. Patricia Hedland, "Virtual Reality Warriors: Native American Culture in Cyberspace," *High Performance* 15 (Spring 1992), 32–33.

34. Frantz Fanon, *Black Skin, White Masks* (New York: Grove Press, 1967), 12.

35. Bukatman, *Terminal Identity,* 213–214.

36. Heim, "The Erotic Ontology of Cyberspace," 80 (italics mine).

Mysteries of the Bioapparatus

Nell Tenhaaf

We live in a thoroughly postnatural environment of information flow and sensory mediation, a world of representational overload. There is an irony in this situation for artists and cultural theorists. Now that we have developed very sophisticated models for how representation works, we are moving into an information environment in which there is to be no representation at all. How we represent ourselves and form subjectivity, in relation to the real world, will no longer be an issue, as simulated entry into parallel digital worlds comes to supersede this relationship. Because the interface between perceiver and perceived is to be ever more transparent and permit a sense of complete immersion in a highly mediated elsewhere, questions of what or who is represented, by whom, for whom, and with what effect are supposed to disappear. I am calling this imagined condition cyberspace. It signals an order of representation that is supposed to be beyond representation; in concert, the imagined condition called cyberspace embeds a wish that subjectivity be shaped in a radically new way.

My project, then, sounds quite conservative: in addressing cyberspace and the altering of my self that it implies, I am particularly interested in developing a continuity with feminist representational analysis of the past two decades, that is, an understanding of how the self is molded sexually and is otherwise differentiated within an overriding symbolic order, through representation. Ideally, one gains with this awareness an ability to affect the symbolic order, to influence the way that it shapes meaning and consciousness. To my mind, it is necessary to understand subjectivity and representation in a continuum with the psychoanalytic model that has pervaded the discourses of the past two decades, in order to plunge into the datasphere without getting lost. This continuum also helps to connect intimately with the effects and meanings of contemporary technoscientific research.

Psychoanalytic discourse has framed the feminine subject within a history of absence, overdetermined by the body and yet not fully present in it. As illustrations of the pertinence of this thesis, biologically there is the doubled body of pregnancy; psychologically, the deep internalization of masculine goals; or politically, the virtual power of the "woman behind every great man." The familiar feminine condition of being both too close to and too distant from the body, constrained by its material reality yet psychically open to penetration, could be said to characterize the experience of being in cyberspace. In parallel, the hypercirculation of representations in the information environment, of art, media, language, science, can be described as an interpenetration between electronically driven fictions and multiple subjectivities. Such unboundedness in the symbolic order is what makes it seem to be postrepresentational.

My thesis for continuity in representational analysis is this: cyberspace implies a feminization of the symbolic order and of subjectivity. I make this claim not as a deconstructive critique but as a slogan for engagement. The immateriality of the cyberspace dataworld reorders the creation and interpretation of representations in ways that are congruent with feminine subjectivity, as we have come to understand it through psychoanalytic models. I am proposing that this loose concept of the "feminine" be taken as a way to decipher the cyberorganism, tracking its departure from the traditional goals of containment and boundedness of any self or ego. The cyberspace subject is a fluid subject, "is a feedback loop"[1] constantly in formation through an interpenetration between self and not-self. In this sense the cybersubject *is* a feminine subject, but not one that is fixed in any psychoanalytically negative position as object of a technological penetration. This subject is no longer determined by the Freudian *psychischer Apparat* because it has its own desiring mechanisms, premised on the inseparability of its biomass from a vastly expanded array of technological, metaphorical apparatuses. To describe the cybersubject, I will use the term *bioapparatus* subject.[2]

A model for "feminization" might ring alarm bells of nostalgia for the affirmative feminism of the sixties and seventies. I do not deny this tendency, although I do not wish to diminish the expanded politics of difference that subsequently redefined the term feminism. Rather than reuniversalize, I want to locate my gendered point of view within a politics of consciousness that is concerned with human commonality, humans as a species engaging in technological interpenetration and struggling with new dimensions of representation.

The Unbounded Body

Subjectivity is formed through edges, and through representability. These reflexive formulations have been central to twentieth-century thought about identity: it is generated and managed through an apparatus that organizes the symbolic around a central core of self, both physical and psychic. An inability of the subject to make an adequate representation for an inner state of being, one that can then be communicated to others, has been described as sickness of the soul, a failure of consciousness, a cognitive error, or neurosis, since Freud renamed the soul as psyche at the turn of the last century and drew it into the sexualized "underworld of psycho-analysis."[3] The difficulty of any "other" caught in the universalizing schemas of the symbolic order built around the psychic apparatus has by now been much discussed. Only an idealized westernized masculine subject has had presence; all else is defined by absence.

In her recent book *Les nouvelles maladies de l'âme,* Julia Kristeva reattaches the idea of soul to the psychic apparatus in order to address the organizational inadequacy of the contemporary psyche confronted with the barrage of representations coming at us in media/information culture.[4] She makes psychic life and soul once again interchangeable, to expand the possibilities for interior life and for meaning in the symbolic domain, because it has become increasingly difficult to integrate the overload of sense perceptions with our inner representations of self. But in the techno-elsewhere of the info world, of cyberspace, it is not that the psychic apparatus can no longer cope; rather, this modeling of the self around a mechanist and masculinist functionality of identity has been completely eroded. Kristeva's query to the reconstituted soul, with its neometaphysical overtones, "are you there?," is more appropriate to the now exploded psychoanalytic terrain. It can absorb the affinity of the subject in cyberspace for dissolution, can accept a desystemizing of psyche and soma that elicits the more fluid and feminine bioapparatus.

One way of approaching this premise of a dispersed apparatus is to consider an old paradox of subjectivity that takes on new proportions in the wired culture of cyberspace. The paradox is this: identity formation or subjectivity involves a process based on visual and textual representation of coming to know oneself. This is Freud's and all of Euro-American culture's formative idiom that comes from Oedipus's confrontation with the Sphinx. It is fundamentally the knowledge of a distinction between the self and the not-self. Yet, at the same time, to contradict or deny this articulation of a familiar, differentiated, and therefore intact self has long been associated with an enhancement of subjectivity, a way toward the perfected self.

This idea can be traced through the history of mythologies and religious beliefs, sexual practices, artistic inspiration, and proto-philosophical discourses such as alchemy, through instances of mystical, metaphysical, erotic, or sacrificial transubstantiation of the body and the self. Consciousness transcends the limits of the biological, gendered, and degenerating body, often through the body being physically or metaphorically opened to and penetrated by the not-self. Such a willed loss of being leads ultimately to a better knowledge of oneself, a stronger subjectivity, a more nearly perfect soul.

Psychoanalysis takes up this theme, beginning with the impact of Freud's theories on his immediate environment of late nineteenth-century subjectivity, that is, the breaking apart of those subjects' coherency already so tenuously held together by will and rationalism. Walter Benjamin has referred to the surrealist project of the 1920s, rooted in psychoanalysis, as the "dialectics of intoxication: In the world's structure, dream loosens individuality like a bad tooth."[5] Subsequently, our century has been consumed by an internalized and externalized struggle with *l'inquiétante étrangeté,* Freud's *unheimlich* that haunts us as the other within, our own strangeness to ourselves in Kristeva's words. Further, without reducing consciousness to an epistemological question, such

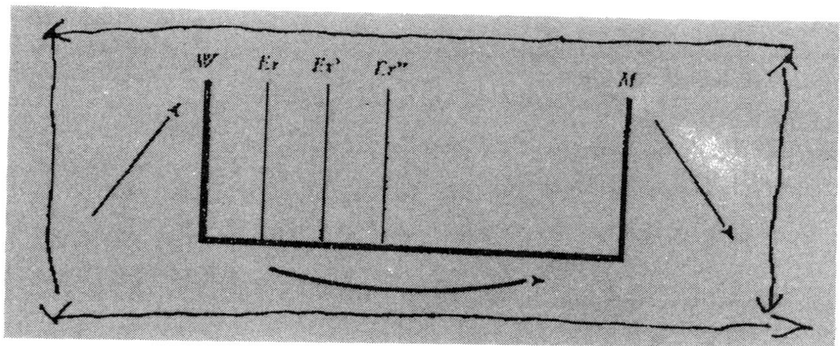

10. Sigmund Freud, formation of a dream, in *Interprétation des Rêves* (Paris: Presses Universitaires de France, 1900).

scarring encounters with the not-self are also philosophical tropes of the way to knowledge, Faustian tales of intellectual curiosity.

The paradoxical conception of a subject who leans toward both resolution and incoherency has strong currency in postmodern thought; it is in fact one way of describing the "subject in crisis" at the core of postmodern thinking. Poststructuralism has construed this anxious subject linguistically, through theorizing its endless deferral in signification (*la différance*). But the idea of a subjectivity that is persistently incomplete and in process is perhaps most comprehensively delineated in the French feminist theory written between the mid-1970s and mid-1980s. The philosophical, psychoanalytic, and linguistic discourses of Kristeva and Luce Irigaray, among others, respond to a historical lack of any recognized subjectivity for women, and so they display a certain audacity in their insistent problematizing and risking of the contemporary female subject.

Kristeva formulates her sense of a provisional subjectivity, a continuing process of separation of self from not-self, as an abjection of repulsive parts of the self begun in infancy.[6] Irigaray has consistently refused any singularity of the subject, any fixed point, returning over and over in her writing to subject/object dissolution through images of fluids and mystical searing of the body. "The wound comes before the flame," in *la mystérique*.[7] Irigaray perceives woman *as* the not-self, she "both is the boundary/the other *against* which identity is constructed, and that which confuses all boundaries."[8] Her project permits a reclaiming of the negative position for feminine subjectivity, as psychoanalytic terminology would frame it. The feminine as lack is a threat to the coherency of the self and must be introjected by the masculine subject through the oedipal scene, as the repressed of his desire. Once it is universalized, defense against otherness comes to constitute a project at the core of the masculine psyche.

The importance of Irigaray's logic is that it subverts the intuitive counterargument to the oedipal scenario, one that is so easily reduced to a double negative: a feminist argument for not-lacking. Her logic for a feminine space of subjectivity is such that, rather than refusing to order the symbolic around threat of and disavowal of lack, it circumvents this order. Irigaray's philosophy encompasses a positive dimension of desiring autonomy—her controversial trope of the two feminine (labial) lips that touch each other in an erotic plurality[9]—but more importantly it embraces as well the lack-of-having, Lacan's *manque à avoir.* It embraces the negative space that admits the not-self.

This is the articulation of a feminine subjectivity of "interpenetration of self with otherness."[10] If women have philosophically and politically reconfigured our condition of political and cultural invisibility over the past decades, there is nonetheless what I would call a valuable feminine knowledge to be retained from the experience of interpenetration, psychic and somatic. It seems pertinent, for instance, to the question of the soul and cyberspace: what is the nature of interiority in a domain where the self is de- and reconstituted in a mode of willed reassembling? The body and psyche are wounded to enter cyberspace in the cyberpunk scenario that William Gibson proposed, and from which the term *cyberspace* entered common parlance. The subject whose imaginary self enters the dispersing space of cyberspace is injured by technoscience—by implants, by self-objectifying for the purposes of experimental enhancement—to the extent that the psyche seems to merge with the intelligence of the net, electronic parts substitute for fleshy parts, and the distinction between self and the cyberenvironment is blurred if not nullified. If it is a fantasized body that is penetrated in cyberpunk fiction, a desire to merge psychically into some kind of cyberworld has nevertheless entered the popular imagination. The widespread fascination with virtual reality technologies attests to this, the wish to experience a perceptual event so immediate that it eliminates the self who must ascribe meaning to it.

This is a condition of technological symbiosis, the psyche penetrated so that the self leaks out and the not-self flows in. Subjectivity has an altered meaning and representation seems to be suspended, because there is a collapse of the familiar paradigm in which self is distinct from other, and subject "reads" object. This issue is not new to feminist theory, in fact it is at the core of the deconstruction versus essentialism debates of the early 1980s. An often cited example is Irigaray's image of women's two lips that "speak together" in a language specific to body, desire, and subjectivity. These lips cannot be reduced to a singular (masculine) entity and thereby generate an excess of meaning that escapes the domination of phallic discourse. In the deconstructivist model, heavily influenced by the school of Lacanian psychoanalysis from which Irigaray had broken in the mid-1970s, such an unmediated relationship between body and language is perceived as an essentialist and even biologistic limitation. It threatens the possibility for female subjectivity because it challenges the conviction that the terms feminine, masculine, body are "textual" constructs in representation, not imminent to the self.[11]

When it is imagined as fully immersive, the experience of interfacing with dataworlds assumes an imminent multiple and "readable" bioapparatus body. It entails its own excess of signification, if only in the complexity and inseparability of somatic and cognitive information to be processed in the human-technology circuit, and this also seems to place the subject in an expanded symbolic system. But there appears to be a complete absence of representation in the immersive cyberspace experience, because it is perceptually so immediate as to seem unmediated. This makes it difficult to pull the excess of meaning into subjectivity in the way that Irigaray might, as enhanced self-representation. Rather than returning more knowledge to the self, hypersignification in cyberspace overthrows representation and projects it into a machine consciousness with a new order of transcendence attached to it. This signals a troubled relationship between psyche and symbolic order in the dynamic of cyberspace, but it could have great potential for enlarging upon the idea of self. The mediation in play between bioapparatus subjectivity and the symbolic order would have to be less dependent on familiar textual structures of signification, and more involved with emerging psychobiological representations of the cybernetic organism.

Politics of Consciousness

Cyberspace, its subjectivities and its power dynamics, has some specific genealogical characteristics. It is derived from Norbert Wiener's now classic cybernetics theory of the 1940s.[12] In cybernetics, the Cartesian idea of body as machine is merged with an informational conception of the body as a self-regulating system. Mechanisms of organization based on coded messaging and computation are derived from both individuals and groups, and these are extended outward into controlling devices or servomechanisms that can be made to simulate and regulate the behavior of an organism or any complex structure through feedback systems. The principle is that laws of communication and of management apply equally to living beings and machines, thus constituting a symbiosis. Wiener himself historicizes these ideas as a reconstitution of the ancient mind/matter debate, from Spinoza's rationalism through Leibniz's all-pervasive vitalist life-force to the nineteenth-century conception of the body as an automaton regulated on the model of a steam engine and, finally, to his own information/communications paradigm.[13]

Wiener's cybernetic model was built on the recognition of a fundamental premise of the quantum mechanics revolution, that is, the play of human indeterminacy between scientific theory and reality, the observer as intrinsic to the observed. The cybernetic human/machine circuit thus seems to offer a technological metaphor for subjectivity that recognizes a complete and unique human entity within the informational loop. But not only did it remain embedded in a Cartesian universalist conviction of mind ruling body, it "appropriated the idea that humans introduced uncertainty into the system and . . . managed to subsume the messiness of the human observer's role under a system of positive math."[14] Since Wiener's time, this perspective has been taken up as the (more or less frustrated) project of artificial intelligence, and is now carried over into cyberspace. The bias is evident in the foregrounding of parts that play the traditionally masculine role of manager and merge best with controlled computation—brain and nervous system, or the mind—while the rest of the body is an imperfect and inconvenient matrix for consciousness that can be externalized as a data space.

Another way of stating the representational dilemma of cyberspace, then, is that the cybernetic informational circuit overpowers the uncertain subject in its perceptual feedback loop. Although the trip into data matrices through wetware interfaces is proposed to be more real than reality, it invokes a struggle to hang onto the knowledge that this space does not engage the whole self; the psyche clings to the memory that this space is a representation, that is, it clings to a memory of the real body and its formulation in physical space.[15] The body is experienced as an image of the body engaged in a deep penetration or a momentary dissolution into space. The experience is intensified by the sense that this projected space has a metaphysical power, it seems to be or is imputed to be a self-sustaining controlling device beyond authorship, a symbolic apparatus outside the self with the capability of ordering representation and constructing the perceiving subject. Rather than a two-way flow, it is an absorption that reconstitutes control from a powerful external source.

Cyberspace thus reflects the power of technoscience. As Sandra Harding describes it, scientific culture works on a conception of ideal knowledge-seeking that excludes itself from what it prescribes for everything else, which is causal analysis and critical scrutiny of inherited beliefs. For Harding, this demonstrates magical or religious thinking in that it assigns a consciousness to scientific and technological progress itself.[16] This kind of mechanism is ideological, but it can manifest itself linguistically in a compelling and impenetrable tautology such as the Darwinian maxim of fitness: "fittest" can only describe organisms that survive, yet predictability is ascribed to it within the premises of today's genetics research. In modernist science, nature is ruled by the laws of physics, which are governed by an ultimate if unknowable principle, by God or by a mystical self-ordering, depending on the belief system. Cyberspace is ruled by a fitness tautology about machine consciousness: the human subject is inadequate or incomplete without cybernetic symbiosis, so the technological node in the circuit must contribute the missing higher attributes. The question is, how does the human agent stay in the picture when technoscience is extolled as a sentient and transcendent power?

The informational web of cyberspace again writes out the subject, in an ambiguous gesture aimed toward both control and surrender, when it is claimed to have a life, an intelligence, and an evolution of its own. A "will to live" is attributed to technology itself via a globalized circuit of "smart" representations composed of digital simulations from biomedical visualization, biotechnological

SHOPPING FOR VALUE (SYSTEMS)
DATABASE DIAGRAM

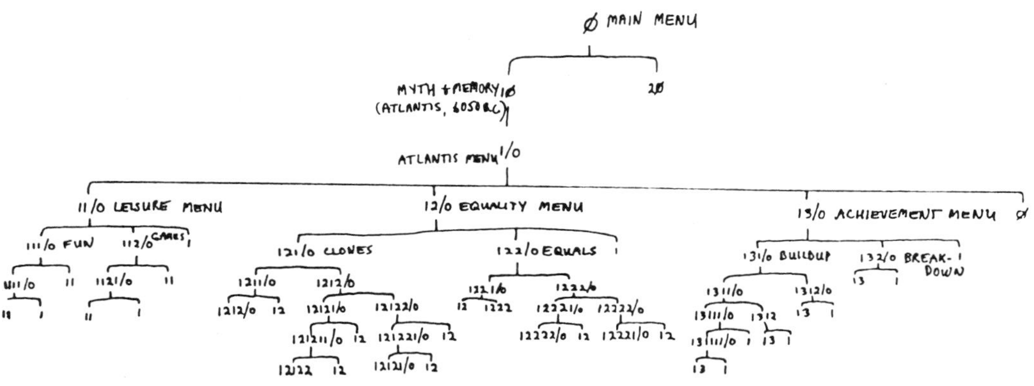

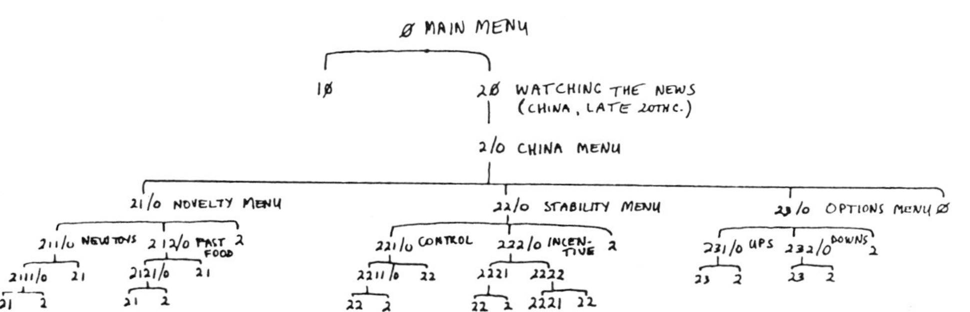

11. Database structuring diagram for a videotex project by Nell Tenhaaf, *Shopping for Value (Systems)* (1984). Photo courtesy the artist.

molecular dynamics, bioengineering dumping into genetic databases and protein banks, neuroanatomy, artificial intelligence, artificial life. Drug-controlled behavior design seems to be on the horizon, and what constitutes public consultation in this domain is managed through the media, inextricable from the goals of corporate culture. Through its current obsession with the medical and ultimately transmutational potential of unlocking the body's secrets, the scientific hierarchy—with physics at its top—reinforces the perception that power ultimately lies in codifying the past, present, and ostensibly the future of the human organism and its environment.

Immersion "in cyberspace" is a revealing metaphor that implies both engagement and escapism. In cyberpunk fantasy, neurological, endocrinal, and other physiological correlates for the body can be tapped in a bio-info conduit, but the objective is rarely one of absorbing technoscience's representational schemas for a gain of self-knowledge so as to engage the power politics of science. Cyberspace ideas of altering the body are enmeshed in technopotential—that is, in what technology is perceived or wished to be capable of doing, generally projecting a virtual body that is artificially powerful and perfectible but submissive to the higher order of the datasphere.

To resolve or eliminate the complex politics of representation emerging in the information order, some purveyors of crisis mentality signal a simple solution, announcing the end of representation altogether. A mundane example that nevertheless has important ramifications is the declaration in the digital art world that simulation by computer processes supersedes old-fashioned representation. As much as contemporary art practices may be overburdened by self-conscious image management, this tired avant-gardism is not a useful contribution. French computer artist and theoretician Edmond Couchot sees the assumption of responsibility by the computer and the computerized image as the automatization of the artist's responsibility for what constitutes art—the *image responsable* takes over. He links this with the end of subjectivity, since choice, chance, fate are no longer issues for desire or will but result from "manipulation of hyper-sophisticated technoscientific *savoir-faire*."[17]

The investigation of consciousness and of other attributes of life through computer systems could engage the politics of consciousness if it were to contribute to the intelligibility of the bioapparatus, by providing a scientific and technological epistemology to complement the psychoanalytic model of subjectivity. This would form a representational complex that recognizes science as a field of desire and intersubjectivity rather than a quest for truth, and cyberspace as a technoscientific fable for loss of self, loss of the historical masculine conception of the intact and solvable "I." It might be germane to conceptualize being in cyberspace as a two-way flow between, on the one hand, a technoscientific organic and intelligent world in simulation and, on the other, the self with its necessary technologies of representation and affect, its self-articulating bioapparatus.

The Fabulous Thinking Body

If my earlier characterization of feminine cyberspace is a *femme* fantasy of fluid subjectivity and an authentic, readable body, femininity itself is clearly a psychoanalytic concept. My understanding of the feminine is filtered through Irigaray and Kristeva. Therefore, it is based in an articulation of the mind/body/psyche complex that has been problematized by psychoanalytic feminism, by its fascination with and fear of the idea that the body *is* feminine. In the now familiar terms of this schema, repressed otherness manifests itself as a symptom that speaks through the female body in a mode of expression that is primary, maternally linked, pre-oedipal, presymbolic. The limitation of this proposal is that, since it is a linguistic model, its portrait of the body is not as a biological entity but as a textual one; or, if the body is seen as biological, apprehension is focused on reductivism and essentialism, not on the complex real-world interventions of, for example, reproductive technologies. Also, in this schema the authentic body cannot be masculine, reinforcing a history in which the male body is implicit and universalized and therefore not discursively present.

Freudian psychoanalysis is credited by even its most stringent critics, Irigaray for example, with offering a previously inaccessible platform for the idea of subject. Now, however, our psychoanalytic models of subjectivity seem limited if still endlessly contentious. From Freudian psychoanalysis we have inherited a problem with being in the present, overdetermined by fatalistic outcomes that are encoded in our deepest somatic and psychic structures.[18] As a result, a basic feature of the psychoanalytic model of the western psyche is that realization of the self is delayed to an indeterminate and perhaps unattainable future. Freud's characterization of femininity as passive was informed by misconceptions, a product of both tenacious social convention and scientific prejudices, in this case a limited understanding of the sexually differentiated movement of the reproductive cells.[19] After a century, we seem about ready to overthrow this stereotype. The subsequent Lacanian proposition of subjectivity formed through the mirror effect, a masculine reflection of the world as other, is a particularly singular and cerebral idea. It is no surprise that its dominance as a model for the construction of identity has fostered a crisis atmosphere around subjectivity. In this schema, the preponderance of the symbolic order, of representation and linguistic signification, constitutes a kind of paralyzing extruded consciousness. The refusal of negative otherness is an equally intense zone of combat. No surprise either, then, are the ceaseless attempts to materialize and to confront these threats in the form of ever more sophisticated technologies, right up to the threshold of technologies like virtual reality that "presuppose the temporary abolition of representations," to paraphrase Monique David-Ménard on Freud and libidinal discharge.[20]

Can the desire to suspend meaning, an outcome of the hypersignification of late twentieth-century life, not coexist with the enhanced reflexivity of knowledge promised by cybertechnologies? I am interested in a parapsychoanalytic model for the cyberspace subject that restores a masculine body to the psychic schema, that admits contemporary technoscience, and that addresses the legacy of endless deferral of a realized self. It should be possible to use powerful representational tools to make accessible and meaningful the isolated bits of specialized information about our bodies and minds with which we are currently bombarded by the media. The virtual body of cyberspace may be a way to self-knowledge, the body read back to itself through its bioapparatus loop. But this cannot come about without an integration of representation and of signifying systems with the enhanced perceptual and informational field of cybertechnologies.

Freud used the phrase "popular conception of the body" in an 1893 paper on hysterical paralysis.[21] He talks about hysterical paralysis of a part of the body as symptom of a "dynamical lesion" in the cortex, thus describing a purely metaphoric injury nevertheless manifested in the body as a felt paralysis. It is a functional injury that does not quite follow the rules of a physiological event, getting some of the features of a genuine motor paralysis wrong. Freud says the hysterical paralysis behaves "as though anatomy did not exist or as though it had no knowledge of it."[22] The "popular conception" is an individual's idea of body, and when one part of this conception is disturbed by a psychical event it reorders the real body through the paralysis that is not quite right.

The intelligibility of the subject through physiological signs in hysteria was one of the earliest premises for Freud's investigations of interdependence between the psychical and the organic. A complex link between operations of the mind and movement in the body came to inform Freud's description of the energistic organizing system he called the mental apparatus, even after he moved away from the strictly biological idea that psychological events occur in specific areas of the brain. Freud was always ambivalent about the reductivism implied in a biological basis for psychology.[23] Although psyche formulates the body as idea and is dependent on language, on representation, it is Lacan who later overpowers Freud's biopsychological model of the psyche with the pervasiveness of signifiers. Everything becomes mediated by language, and the body is always an image of the body, a fantasized body. Freud's model for consciousness with all its limitations remains an indexing system, a negotiation of terrain, of inside and outside, of instinctual effects in relation to thinking, libidinal pleasure and unpleasure in complex interaction with representation. The somatic symptom of movement or posture in hysteria is not a Cartesian "radical elsewhere of thought,"[24] but an autonomous articulation related to

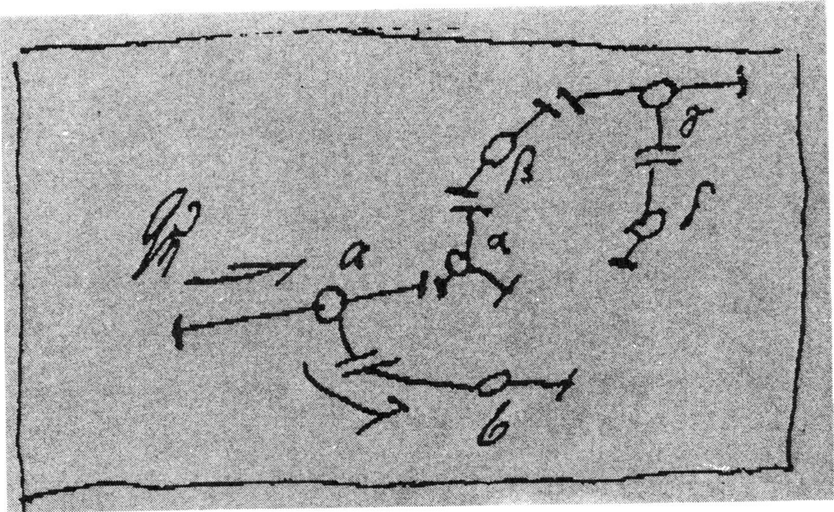

12. Sigmund Freud, preliminary notebook for *Project for a Scientific Psychology*, schematic representation of the primary defense (repression), reproduced in *La Naissance de la psychologie* (Paris: Presses Universitaires de France, 1916).

thinking that is, for reasons of social malaise, inarticulable in language. Although his reasons were neurological, Freud himself classed hysterical paralysis as a "representation paralysis."[25]

Freud's writing tracks the intellectual and scientific shifts he makes between the many facets of the self, and this is true of his construction of organicist metaphors in which complexity overrides biological reductivism. In parallel, the science that I am interested in mapping into bioapparatus subjectivity is not simply another representational imperative driven by the linguistic order, collapsing nature into fiction or the body into the hyperreal domain of simulation. It is a science of negotiation between different registers: the natural world; narratives of evolution; pharmaceuticals and power politics; aesthetics of protein molecules; projected subjectivities in human/machine links. This science, the science of the bioapparatus, returns to the Freudian epistemology of the "body that thinks,"[26] which not only cannot admit a mind/body dualism but insists on a mysterious corporeal and representational dynamic beyond the limits of any single theory. To borrow a cognitive loop from Jean-François Lyotard, "Thought is inseparable from the phenomenological body: although gendered body is separated from thought, and launches thought."[27]

While the scientific mapping of hysteria a century ago identified but also pathologized the mute feminine subject, the science of the bioapparatus would permit the thinking body to speak through a compound representational feedback loop: interactive, multisensory, reflexive, knowledge-based. Today's popular conception of the body makes of it a readable, mutable, and manageable information bank constrained by its macromolecular expression, but research into, for example, psychosomatic interpretations of illness reminds us how little is known about mind and body interaction. A close interdependence between the body's nervous system and immune system is now postulated, suggesting that reasons for the mysterious efficacy of placebos in curing disease may come to light.[28] The body seems always to know itself in a mutely expressive way that may by definition not be conscious, but that may nevertheless become more accessible. If cyberspace is read as a "hysterical" fantasy of the late twentieth century, it is expressing an inutterable desire for a more fluid subjectivity and a body that knows itself better, and this by a simulation through technologies of enhanced organic behavior. The idea of body is already wired into a vastly expanded bank of physiological knowledge. That is how up-to-the-minute information and imagery about the body's deepest workings and its most complicated biosocial functions, reproduction and evolution of the species for example, now enter and become interwoven into the biomedical readability of the self.

Merging psychoanalysis and science in cyberspace does not guarantee a better understanding of the repressed narratives of technoscience or its emerging feminine features. It is a way, though, to respond to the objectification of my own subjectivity, to grant it some self-evidence in the face of many complex mediating and distancing discourses circulating out there. I would like fables of the unbounded body to abound in cyberspace.

Notes

Thanks to Eliot Handelman, Céline Surprenant, Inez van der Spek, Marjorie Beaucage, Katerina Thomadaki, and Maria Klonaris.

1. This was Eliot Handelman's response to my first description of the cyberspace subject as embodying a "position on a psychic feedback circuit," rather than embodying the loop itself. For Handelman, "*Loop* suggests something like a functionally bounded region, whereas *circuit* suggests bounded functionality that emphasizes movement and trajectory."

2. See Catherine Richards and Nell Tenhaaf, eds., *Virtual Seminar on the Bioapparatus* (Banff: Banff Centre for the Arts, 1991).

3. Sigmund Freud, *On the History of the Psycho-Analytic Movement,* trans. Joan Rivière (New York: W. W. Norton, 1914; reprinted 1966), 66.

4. Julia Kristeva, *Les nouvelles maladies de l'âme* (Paris: Fayard, 1993).

5. Walter Benjamin, "Surrealism," in *Reflections,* trans. Edmund Jephcott (New York: Schocken Books, 1986), 179.

6. Julia Kristeva, *Pouvoirs de l'horreur. Essai sur l'abjection* (Paris: Editions du Seuil, 1980).

7. Luce Irigary, "La Mystérique," in *Speculum of the Other Woman,* trans. Gillian C. Gill (Ithaca: Cornell University Press, 1985), 191–202.

8. Christine Battersby, "Her Body/Her Boundaries: Gender and the Metaphysics of Containment," in *Journal of Philosophy and the Visual Arts: The Body,* ed. Andrew Benjamin (London: The Academy Group, 1993), 34.

9. Luce Irigaray, "When Our Lips Speak Together," *This Sex Which Is Not One,* trans. Catherine Porter (Ithaca: Cornell University Press, 1985), 205–222.

10. Battersby, "Her Body/Her Boundaries," 34.

11. For example, in Jacqueline Rose, "Dora: Fragment of an Analysis," in *In Dora's Case: Freud—Hysteria—Feminism,* ed. Charles Bernheimer and Claire Kahane (New York: Columbia University Press, 1985), 140.

12. Norbert Wiener, *Cybernetics: Or Control and Communication in the Animal and the Machine* (New York and Paris: Wiley and Hermann et Cie, 1948). One of the sources for Wiener's theory was his work on weapons guidance systems in the Second World War.

13. Ibid., 51–54.

14. David Porush, "Reading in the Servo-Mechanical Loop," *Discourse: Journal for Theoretical Studies in Media and Culture* 9 (Spring-Summer 1987), 55.

15. See for example William Gibson, *Burning Chrome* (London: Grafton Books, 1988), 205.

16. Sandra Harding, *The Science Question in Feminism* (Ithaca and London: Cornell University Press, 1986), 36.

17. Edmond Couchot, "Esthétique de la simulation. Une responsabilité assistée?" *Art Press* 12 (1991), 149.

18. Irigaray, *This Sex Which Is Not One,* 72. Jean-François Lyotard reiterates such a tone of fast-forward into the future in "Can Thought Go On without a Body?," in *The Inhuman: Reflections on Time,* trans. Geoffrey Bennington and Rachel Bowlby (Stanford: Stanford University Press, 1991). See also Lyotard in *Virtual Seminar on the Bioapparatus.*

19. See Sigmund Freud, *Some Psychological Consequences of the Anatomical Distinction between the Sexes,* in Freud, *Standard Edition,* vol. 19, trans. James Strachey (London: Hogarth Press, 1953).

20. Monique David-Ménard, *Hysteria from Freud to Lacan: Body and Language in Psychoanalysis,* trans. Catherine Porter (Ithaca: Cornell University Press, 1989), 8.

21. Sigmund Freud, *Some Points for a Comparative Study of Organic and Hysterical Motor Paralysis,* in Freud, *Standard Edition* vol. 3, trans. James Strachey (London: Hogarth Press, 1953), 170.

22. Ibid., 169. Hysteria simulates organic nerve disorders, and in the instance of hysterical paralysis it mimics the characteristics of a paralysis *en masse* that usually results from a cerebral lesion. It differs especially from organic paralysis in its "precise limitation and excessive intensity."

23. Frank Sulloway, *Freud, Biologist of the Mind: Beyond the Psychoanalytic Legend* (New York: Basic Books, 1979); see the chapter on "Freud as Crypto-Biologist: The Politics of Scientific Independence."

24. David-Ménard, *Hysteria from Freud to Lacan,* 141.

25. Freud, *Some Points for a Comparative Study of Organic and Hysterical Motor Paralysis,* 162.

26. David-Ménard, *Hysteria from Freud to Lacan,* 12.

27. Jean-François Lyotard, *The Inhuman: Reflections on Time,* trans. Geoffrey Bennington and Rachel Bowlby (Stanford: Stanford University Press, 1991), 23.

28. Gerald Messadié, "Placebo . . . Le cerveau qui guérit," *Science et Vie* 920 (May 1994), 64–69.

When Is the Ear Pierced? The Clashes of Sound, Technology, and Cyberculture

Frances Dyson

The discourse and technology of virtual reality (VR) entail a redefinition of embodiment, space, reality, community, and authentic experience—all concepts that have already undergone certain modulations in the discourse and practice of audiophony. Indeed, it could be said that everything that happens in VR technology is no more than an accumulation of the auditive technologies of the past: a realization of the telepresence and interactivity first offered by telephony, a computation of the inscriptive strategies of the phonograph and tape recorder, an appropriation of the disembodied "presence" of radio, an embrace of film sound's spatiality, and an instantiation of the hyperreal sound-effect present in all auditive media.

It could also be said that, metaphysically, cyberspace has already been filled with sound, and VR is already an "aural" medium. The features that differentiate VR from other media, that boast that one can enter the screen, can become immersed in and interact with three-dimensional images or "virtual objects," can possess a subjectivity simultaneously embodied and disembodied, can enjoy authentic rather than mediated experiences in a new "virtual" community, and can witness the emergence of a new "metaphysics of the immaterial"—these features are "there" in the *Dasein* of sound.[1] However, the rhetorical maneuvers that align virtuality with aurality, that tell us that we can enter a space of no space with a body that is elsewhere, conflict at the deepest level with the ocularcentric, abstract, and ultimately disembodied basis of VR's technomythology. As much as it provides a site of immersion, aurality also exposes the grain of these contradictions.

Into this net of configurations virtual audio itself plunges. Moored in the floating spaces of telephony and radio, enveloped by the neoorganic and neoromantic discourse of the "global village," mapped by the stylus of phonography and tape recording, and now ejected into the semiotic ocean of digital sampling, virtual audio incorporates multiple dimensions. But it does so in a way that is by far an advance on other audio media because, in the pursuit of immersion and embodiment that is VR's stated goal, virtual audio actually takes the body into account. In fact, virtual audio acknowledges and technologizes the presence of the body and the environment in a manner that the visual aspect of VR is only just beginning to encompass. Because of the confluences between virtuality and aurality, there is a very good case for reading the culture and technology of VR in terms of sound. This paper attempts to indicate some of the discursive and mediamatic consequences of such a reading.

Aurality and Audiophony

Western epistemology has, since the early Greeks, invested as much energy in keeping sound quiet as it has in making matter mute. The phenomenal invisibility, intangibility, multiplicity, and existential flux of sound challenges an ontology grounded in the visible, material, singular, and stable object. Because sound is located at the site/sight of the object, as Christian Metz was first to point out, it is considered to be an attribute of objects, an "infra-object," perhaps one could say a virtual object.[2] The aural object is intrinsically oxymoronic: relying on onomatopoeic words like *crash* and *thud* in order to be named and represented, it oscillates disturbingly between poetry and babble. The perpetually incomplete representation of sound is reinforced by the subjectivity of the listener, who must shift from the dominant perspective of the look—always *in front of* the subject—to the more peripheral, enveloping process of hearing. While the 180-degree range of the eyes position the subject as director of his or her look, so that "looking ahead," projects them toward the future, the 360-degree range of the ears subverts this role. An always open orifice, the ears allow the subject to be continuously and uncontrollably surrounded by sonic disturbances, refusing to limit what is behind them, what is past. Within the sonophobic regime of the West, sound further attacks the integrity of the subject by ignoring the boundary of the skin. Present both externally, in the environment, and internally as a resonance or vibration, sound erodes the limits of the culturally well-established body. This onslaught is leveled at the very logic of conceptual construction of the western body, for in creating a notion of interiority as an avenue to the "truth" of the body and the spirit, sound at the same time destroys the possibility of distinguishing between subject and environment, self and other, interior and exterior. Immersed in sound, the subject thus loses its self.

Throughout the course of media history these aural characteristics have paralleled notions of electronic transmission. Thus it is not surprising to find the enthusiasm surrounding the discovery of telephony and wirelessness at the beginning of this century—expressed in terms of the immateriality, fluidity, lack of mediation, and cosmic "truth" of the ether—resounding through the new electronic frontier emerging at the century's end.[3] Nor is it surprising to find the current excitement around VR following the pattern of earlier developments. Telephony itself arrived at a time when electricity was closely identified with a spiritual and/or cosmic force,[4] and electronic transmission with the movement of the soul through worldly and heavenly spheres. These associations became particularly compelling given that the first material of electronic transmission was the human voice, which, as a major figure in the field of telecolonization, spread western speech with missionary zeal across the earth just as broadcasting, originally an agricultural term, spread the seed of the future crop.[5] The notion of spreading the seed has, of course, strong biblical connotations, uniting the electronically transmitted voice with the invisible voice of the Judaic and Christian god on the one hand and the inner voice of western metaphysics on the other. As an instance of proto-telephony, quattrocento paintings also represented the transmissional device carrying the word/seed of God as a giant speaking tube connecting the deity's mouth with the Madonna's ear, while in the vernacular of nineteenth-century culture, mediumship, or "voice boxing," was also known as "the wireless telephone."[6]

The subjectivity represented by the transmitted voice is thus charged, grounded in the characteristics of the unified, reflective, and self-conscious ego of the Cartesian subject, while also transubstantiated through an electrical disembodiment and dispersion. Caught up in some placeless communications network and vulnerable to interference, crossed lines, and abrupt terminations, the telephonic voice is foisted into what could be thought of as a proto-cyberspace, filled with the appropriate phantoms of the time, just as contemporary cyberspace is filled with "lines of light" and "memory constructs." As a reading of a recent VR artwork in this essay will show, both versions of cyberspace share a fascination with the ephemeral pathways connecting agency and corporeality.

With the advent of radio and the very idea of wirelessness, the lines connecting the transmitted voice to the speaking body were cut. Initially a two-way medium, radio was heralded as a means for linking a multitude of minds through signals originating in spiritualized ether common to humanity. In this sense, it shared the ideas of a democratic purpose and utopian design surrounding cyberspace today. So great was the rhetorical power of radio that it became a resource for many of the grand modernist reconciliations, uniting the listener's ear with the social body and the heavenly bodies alike.[7] This integration was seen as a product of both the technological/ethereal space made accessible by radio and the essentially *aural* nature of the medium. Walter Ong's concept of secondary orality, for instance, offered a potential means of delivery, both *of* information and entertainment and *from* the anxieties wrought by industrialism and urbanization, through the ideals of community, integration, and tradition found in preliterate *oral* cultures.[8]

Gaston Bachelard's radiotopia jettisoned the ether-made-audible in a bid for access to the unconscious. In "Reverie and Radio" he locates radio within the *logosphere,* a sphere in which the individual unconscious can communicate with others and by doing so articulate "a certain universality."[9] Radio broadcasts to the interior of the individual—the "house of our dreams"—within the interior of our home—"the seat of privacy and inwardness."[10] A critical component of such privacy is the anonymity of the listener; indeed, "this lack of a

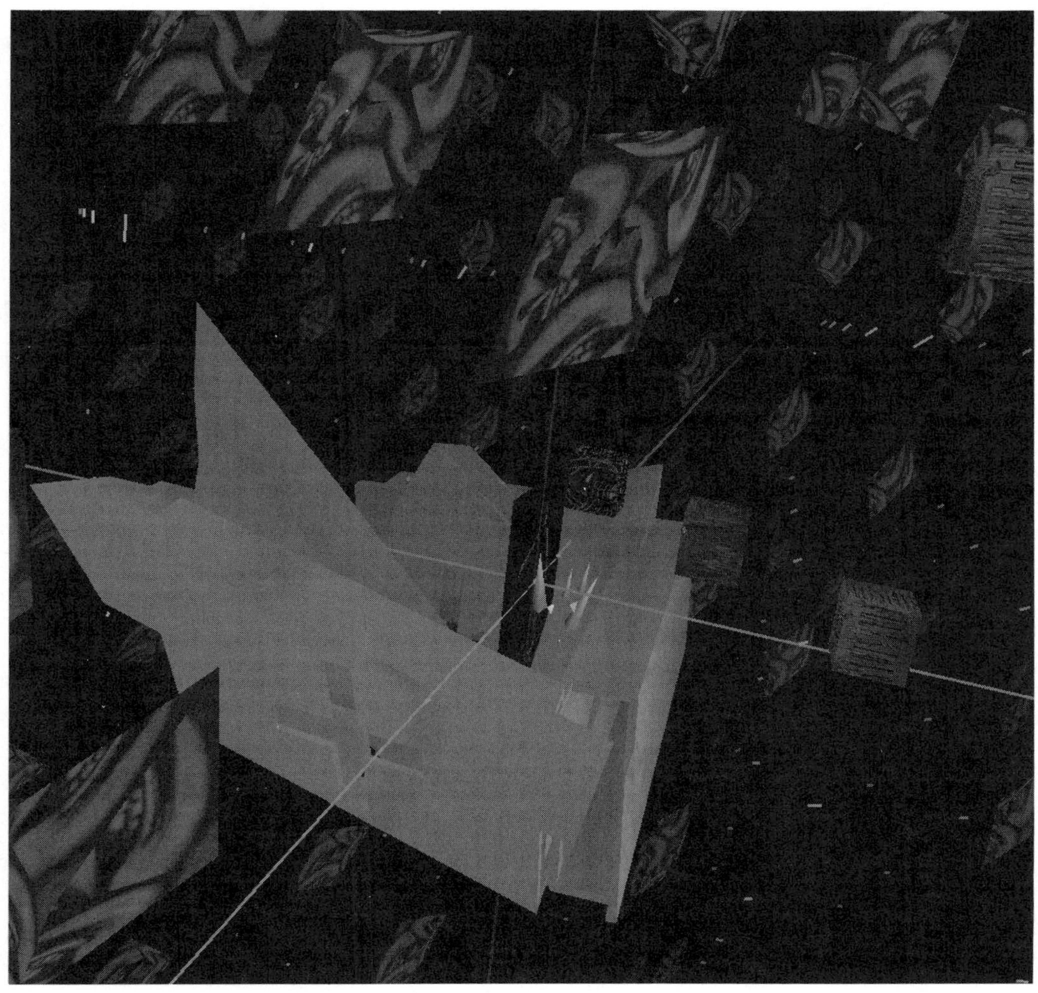

13. Marcos Novak, *Dancing with the Virtual Dervish: Worlds in Progress* (1994), frame grab of one of four virtual worlds. This one shows the maze chamber. Digital image courtesy the artist.

face to face to go with the voice is no impediment; rather it is an asset because it is precisely this which opens up the axis of intimacy, the inward perspective."[11] In cyberspace such anonymity allows access to an individualism represented less by the creation of a singular inward perspective than by the multiple selves the user chooses to broadcast.

Rudolf Arnheim linked the interiority and intimacy of radio to the technologies of broadcast, arguing, for instance, that the proximity to sound offered by the microphone gave the listening experience a particular kind of authenticity. The intimate radio voice creates, in Arnheim's words, a *Stimmung* or "atmosphere" associated with the "cozy parlor" on one hand and the "Heavenly Father . . . unseen yet entirely earthy" on the other.[12] In the metaphorically saturated space of radio, the microphone has a special relationship to the sacred: "The physical fact that the normal distance between sound-source and microphone is *inconsiderable,* implies as a normal condition of the art of broadcasting a *spiritual* and *atmospheric* nearness of broadcaster and listener."[13] In Arnheim's schema of proximities the spiritist origins of radiophony, the broadcast apparatus, and the listener are now ontologically linked: "listening to" is no different from "being there." But the crucial question now becomes: what is this there? is it aural or mediamatic?

Eradicating physical distance in order to deny a metaphysical difference—between the real and the representation—is a strategy that also operates in VR. Just as the headphones concentrate the articulation of space in the ears, the VR "eyephones" (as they are called) concentrate the visual field directly in front of the eyes. There is as much space between headphone and ear as there is between eyephones and eye, and this physical fact also seems to glue the notion of reality to the computer simulation. The "reality" of VR, then, has less to do with overcoming the problem of distance figured in centimeters than with overcoming the ontology these few centimeters of distance represent.

Computer animation artists Monika Fleischmann and Wolfgang Strauss sketch the parameters of this ficto-physical space. By enabling the user to fly, to go through walls, and to see objects from the inside, for example, VR technology is, they claim, capable of creating a new space of perception and embodiment. This embodiment entails a new definition of being: because "the viewer stands *in*—no longer merely *in front of* the picture," Fleischmann and Strauss conclude that being "in" the picture means being part of the picture to the extent that one can say "I am in the picture—therefore I am."[14] If this new existential certitude is based on the elision of physical distance within an essentially frontal perspective—the cybernaught still looks into the screen—we need to ask where the power of this "in" lies.

As a way of representing the body in space, according to a perspective inherited from the Renaissance, frontality is associated with futurity and opposed to anteriority. The stereophonic perspective of traditional audio is essentially a frontal view, incapable of admitting any "behind" into the acoustic scene. The listener is absorbed into this frontality: with headphones creating a sense of intimacy and interiority, with electronic transmission providing a direct line to the future, and with stereophony delimiting an auditive range modeled on vision, an embodiment and phenomenality is established that VR is able, literally, to step into. As the cybernaut looks in front, to the screen, or in/to the helmet clasping the eyes and, from there, in/to the scene where the virtual hand or cyberbody hovers in front of the eye, she or he enters "into" cyberspace and, in doing so, enacts a ritual of progress. The cybernaut need never look back, need never think of this progress or projection as perhaps a retreat—from face-to-face sociality, from the spaces populated by bodies, environments, other species, from the dark matter, or "meat," residing below the neck.

Virtual Audio Reality

Like the visual component of VR, virtual audio technology creates a three-dimensional, interactive aural space or "sound field," in which sounds are heard as if they occurred in the lived environment. But unlike preceding audio media, virtual audio is entirely a product of the digital age, and this has important ramifications. In digital technology the indexical trace of the phonographic and filmic surface, which can be haptically manipulated, is replaced by the arbitrary code of the signal, and thus has no necessary relation to a singular body or an actual sound. While digital media implicitly define sound in terms of a "thing," that "thing" is no longer an object (of the concrete kind) but a stream of information constituted by zero and one values that are *approximations* of discrete samples of audio waves. These approximations simulate the analog sound wave, reproducing characteristics that the listener will perceive to be "true" of the sound rather than the actual characteristics of the sound itself.[15]

Representing the sound wave as a series of discrete particles that can be infinitely divided, arranged, and combined, digital audio transforms sound into a nano-object—an object capable of being manipulated at subsonic, inaudible durations (less than 1/20 of a second), whose manipulation is therefore undetectable. The impossibility of hearing the digital edit with the human, organic ear calls into question the ear's usefulness vis-à-vis this technology. Indeed, within the telematic network, all bodies disappear: the body of the sounding instrument is replaced by the MIDI array and sample bank, the body of sound is replaced by data, the body of the indexical trace is evacuated, leaving only the shell of a sign. As if rearticulating the visualism of metaphysics, sound is furthermore represented by the image of a sound wave.[16] The ear of the listener is thus transformed into a *hearing eye*.

Developments in virtual audio, however, threaten to shift this visual orientation, providing a schematic that actually unravels the logocentric grid as it supposedly amplifies it. It does this through two main mechanisms. First, unlike previous audio technologies, which offer at best a stereophonic perspective—a "hearing ahead"—virtual audio offers a 360-degree aural space. Second, this three-dimensional space simulates the actual movement of sound in an environment, mapping sound's passage from a virtual source through a particular acoustic environment to the ear, which, it must be stressed, is no longer abstracted as a hearing eye but, rather, is represented in its full fleshy uniqueness. For computer technology to admit the listening body into the interface is quite radical. To admit the sound source as an independent, autonomous identity is even more so.[17] Despite the fact that these acknowledgments reflect a pragmatic rather than a metaphysical shift, their effect is to reconfigure the parameters of listening.

The kind of mapping necessary to both locate sounds in space and replicate the physiology of auditive processes is immense. The virtual sound source must be described according to its distance, elevation, azimuth, and environmental context (the effect of the room on the sounding source). This perspective must then be calculated for each and every sound in the sound environment, and has to accompany each sound as it moves or "transmutes" over distances and through space. If static, the sound source should seem to remain constant despite movement of the user's head. Reverb (for the room environment) must be calculated for its crucial role as a cue for the localization of sound, and this means calculating reflections from walls and every other structure in the environment for each sound.[18] These factors are extremely important in simulating an immersive space, since the "there" of an environment is registered in minute detail by the ears. With sound there is no "nothing" between objects, no volume that can be ignored. Every object, every form affects the resulting sound, every surface becomes a topology that yields reflections and, by registering the sound absorbency of the materials, reveals a depth. In the virtual environment, sound both travels through and bounces off objects. In doing so it marks materiality and immateriality at the same time, representing the world as matter on the one hand, as a continuum of differing pulses and transient formations on the other.

Enveloped by this perspective, the ear, or rather the body, of the listener, together with the mechanics of hearing, undergo a similar redefinition. First, the "interaural-amplitude differences and interaural-time differences" that are produced by having two ears separated by some distance have to be calculated to simulate binaurality.[19] Second, and far more complex, the "head-related transfer function" (HRTF), that is, the individual formations of the ear, the head, and the torso, must be computed, since differences will change the way sound enters the ear.[20] The HRTF measurement represents a major change in the understanding of hearing. Philosophically, it realigns the ears with the body rather than the head, dispersing audition through regions of the body, such as the paunch, that have been regarded as irrelevant to sensory perception and the formation of knowledge. Technologically, it signals the inadequacy of audio media that use a stereophonic perspective to deliver "perfect" sound. Culturally, it introduces a degree of technological intervention in the listening space that, since the time of phonography, has been considered private, intimate, and individual. This intervention takes the form of a measurement standard that idealizes certain ear shapes; thus researchers will talk of good or bad pinna with reference to the pinnas' ability to localize virtual sound. As Begault writes: "In the future, in fact, I think we'll see musicians and recording engineers carrying around special sets of data for ears, the same way keyboard players carry around synthesizer patches now. . . . On the backs of CD packages you'll see credits like, 'This CD was recorded with Michael Jackson's ears.' "[21]

Sound Effects

Nonetheless, the problem of "bad pinna" persists and, as researchers admit, it cannot be overcome simply by transplanting Michael Jackson's ears (for example), onto the average audiophile. "Reading" virtual audio is an acquired skill; it must be learned in the same way that audiophiles have learned to "read" telephonic, radiophonic, phonographic, and cinematic sound. The localization of virtual audio thus begs the question of what kinds of sounds were actually audible in previous media, and what kinds of ears were listening. It is important to note for instance that the HRTF "earprint" represents the journey of sonic signals as they are transmitted through loudspeakers in an anechoic chamber. An anechoic chamber is of course devoid of any sound, so the original sonic event, the event that in analog recording may have been in the field, is already at one remove. This map is then reproduced through headphones, detaching it even more from the "there" of the world. Indeed one of the problems associated with localization is that of hearing sound "inside" the head rather than to the right, left, or back. Hearing sound "inside" the head of course defeats the purpose of sound spatialization, since there *is* no space inside the head. Like the interiority of the inner voice, the inner sound thus threatens to make audible a solipsism that is conceptually antithetical to both virtuality and aurality.[22]

This technically abstracted, anechoically derived sound is then placed within an audio-visual context, the closest analog for which is realist cinema. It is fairly well recognized that, as in cinema, the use of sound increases the realism or, rather, the "suspension of disbelief" in the virtual environment,[23] and that high-quality audio makes people perceive visual displays to have higher resolution, while the converse is not true.[24] However, as John Belton points out, cinema sound is more hyperreal than real: "The soundtrack corresponds not, like the image track, directly to 'objective reality,' but rather to a secondary representation of it. That is, to the images that, in turn, guarantee the objectivity of the sounds. What the soundtrack seeks to duplicate is the sound of an image, not that of the world."[25] Duplicating the sound of an image gives cinema sound a degree of latitude in terms of what it can represent; however, it also has the tendency, one that virtual audio has inherited, to reduce sounds to easily recognizable "sound effects." Because, as researchers point out, it is easier to localize the sound of a plane "overhead" than the sound of a piano flying overhead, sounds have to be both recognizable and "in their right place," so to speak.

If the identity and trajectory of the virtual sound must be regulated, so too must the involvement of the listener. Linda Jacobson claims that "all the [three-dimensional audio] systems work in different ways but share a goal: to trick the brain. They fool us into perceiving that the sound is coming somewhere other than the speakers (or headphones)."[26] This "ventriloquism" extends the cinematic practice of placing speakers behind the screen, simulating the unity of sound and image with their respective and often opposing epistemologies. Correspondingly, the synchronization of voice and moving lips (lip sync) reproduces a phonocentric unity of the viewing subject. The persuasiveness of lip sync can in part be attributed to the near impossibility of imagining the voice and body on the screen as separate when the viewer feels their own voice to be "embodied." VR shares this insinuation of the body, which is reinscribed in the virtual data complex each time the head turns, the arm moves, or the eyes flicker. Thus sound and vision seem, if not unified in an object, at least to belong to the same virtual environment in which the body is immersed. The environment, then, takes the place of the phantasmatic body, with the sound contributing its own "aura" of three-dimensionality, spatiality, and the sense of "being in" to the overall effect. Because the cybernaut's body is supposedly "in this environment," the inability to experience this environment as "true" or "authentic" could only correspond to a doubting of one's existence, to a being there and not there at the same time.

Through these technometaphysical processes the cybernaut's body becomes the map upon which the "embodiment" of virtuality is verified and the "presence" lost in the simulation is restored. However, it would be a mistake to dismiss these processes as symptomatic of subjectivity in the age of the hyperreal. Whereas hyperreality still implies some connection, regardless of how faint, to the ethos of verisimilitude, sound has no such loyalty; after all, where there is no "thing" to represent, there can be no "misrepresentation." Similarly, VR, as a space of computer-generated simulation, renders irrelevant questions of verisimilitude, realism, and authenticity. Unlike the camera, the simulation makes no claims to reproduce reality, and in that sense it cannot be wrong, it can only be bad. Without the anchor of realism, a continuing tension develops between the necessity of the body (materiality) and the arbitrariness of the signal (virtuality), between two-dimensional images and three-dimensional sound, between here-and-now plenitude and deference and lack. In accepting the synthesis of these oppositional terms as "real," as belonging to the same ontological category operating within the same actual environment, the cybernaut becomes not a "screen" but an "effect"—arbitrary, abstracted, and reduced.

If VR is thought of as a telephone, its rhetoric is fairly coherent. However, when aurality is used as supplement, both materially and metaphorically, in the creation of an immersive space, the rhetoric of VR begins to stumble. Although the cybernaut shares with the listener an undecided, disembodied subjectivity, a roving kinesthesia and an overlapping sensorium, although she or he exists in a place, like sound, where "the whole world is your body—equally—and

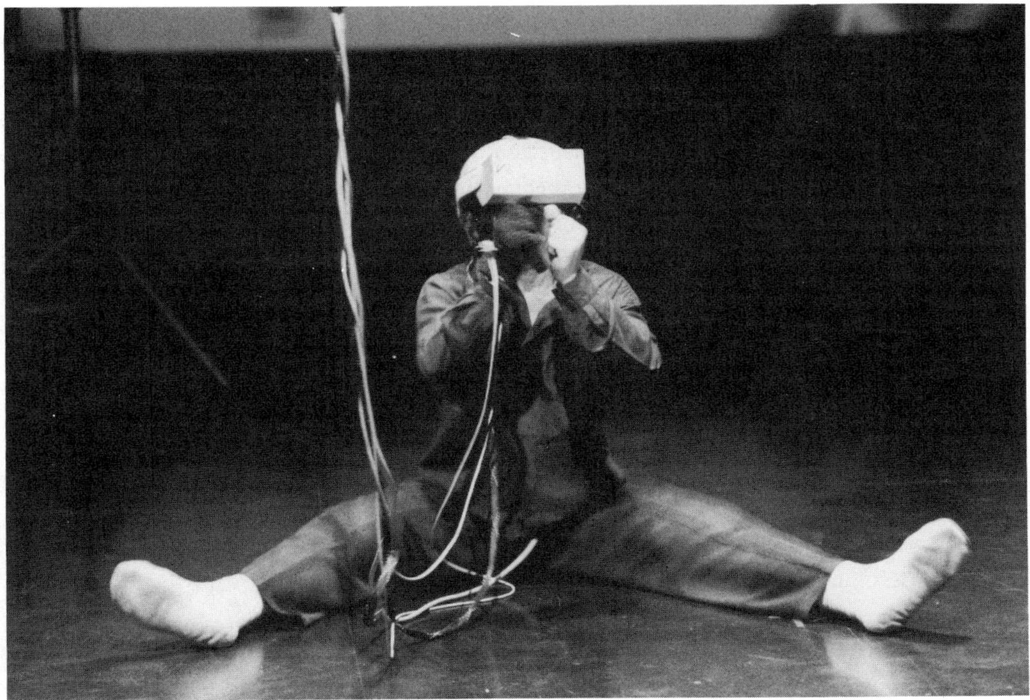

14. Diane J. Gromala and Yacov Sharir, *Dancing with the Virtual Dervish: Virtual Bodies* (1994), performance detail with Yacov Sharir in foreground and virtual worlds projected in background. Photo Cheryl Bellows.

everybody shares the same body,"[27] every point in this fluid, polymorphous space is precisely mapped, both technologically, through computation processes, and culturally, through processes of signification. Although the addition of sound may increase the realism of the virtual event, this "realism" is born not from "the real"—the *hic et nunc* central to the immersive experience—but from "the real" as constructed by other media, chiefly cinema and television. If "you cannot tell the difference between the original and the copy; nor the difference between the copy and a computer-synthesized version,"[28] that may be because the original has been virtualized, while terms like *reality, in, distance, in front of, localize, experience,* or *enter* have meaning only through recourse to the "ground" of an increasingly abstracted corporeality and increasingly fragmented subjectivity.[29] Far more than cinema or television, VR transforms its user, and, as virtual audio makes clear, this transformation involves mapping, dissecting, and computing the body in a manner that heralds a liberation from the dominance of vision yet threatens to colonize the other senses with the same logocentric processes that produced post-Enlightenment vision in the first place. Instead of "becoming what we see," we may turn from the cinema screen to the VR apparatus only to find ourselves "becoming what we be."

Dancing with the Virtual Dervish

Measuring the skin between the body and its environment, representing the body's proprioceptivity by numbers and figures, mapping the minute while invoking the cosmic, all these operations involve a reconceptualization of space which, paradoxically, renders terms like inside and outside meaningless. What would it be like to enter and interact with a "space" of numbers? Could you call it a "space"? Could you call a constellation of data an object or even an image? Could you call a bionic interface manipulating numbers an artist, or would writer be a more appropriate description? Friedrich Kittler suggests that "writing can store only writing, no more, no less. . . . In order to optimize writing for machines, it must no longer be dreamt of as an expression of individuals or as a trace of bodies."[30] The practice of "writing with sound," enabling the aural to be collected in a material object, played backward, slowed down, cut up, spliced together, and reassembled has, since the time of phonography, problematized audio's status as a *nonmediated* medium. Recorded sound cannot claim the so-called authenticity of direct, live transmission, since the recording is tied to neither the here nor the now of the sonic event but rather to a system of representation guided by technology. The inscription of sound thus presents a troubling moment in the discourse of audiophony. Like the shift that suddenly transforms noise into music, glossalalia into poetry, or a cry into language, one experiences an uncanny moment of uncertainty; unlike the Enlightenment ideal of *poesis,* however, this act of bringing forth is also an act of enframing. As aurality and the organic whole it once represented is replaced by audiophony, the possibility of a "source" beyond the field of inscription diminishes in the twilight of the century. No wonder, then, that audiophonic inscription is often represented as a kind of tragic cruelty.

Diane Gromala, Yacov Sharir, and Marcos Novak's two-part virtual environment artwork *Dancing with the Virtual Dervish,* produced during the Art and Virtual Environments project at the Banff Centre for the Arts, powerfully represents the processes of abstraction and disembodiment that constitute the virtual audiophonic experience. In her presentation of part one of the work, *Dancing with the Virtual Dervish: Virtual Bodies,* Diane Gromala spoke of two central artistic concerns: first, to articulate the experience of witnessing her body as a site of medical representation, negotiation, and intervention; and second, to articulate the kinds of strategies she developed to deal with intense physical pain. In both cases, Gromala was concerned with disembodiment—as a transcendent, spiritual state (like that of the dervish), as a function of virtuality, and as a fragmented product of medical culture. As Gromala points out, however, disembodiment is a complex issue in the realm of VR, since the disembodied feeling of being "in" the virtual environment is countered by the grounding imposed by the body, gravity, and the physical limitations the "gear."[31] Disembodiment becomes even more complex in this specific artwork because the virtual environment is a pseudo-realistic rendering of the body's interior.

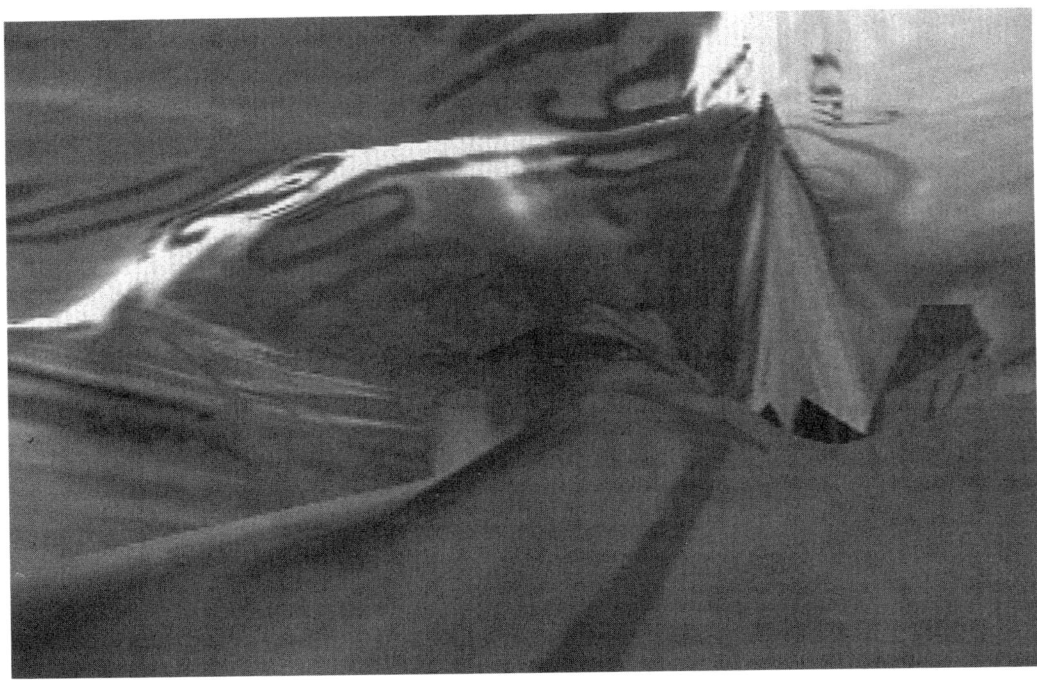

15. Diane J. Gromala and Yacov Sharir, *Dancing with the Virtual Dervish: Virtual Bodies* (1994), frame grab of texts within the environment. Digital image courtesy the artists.

Quietly grotesque, this interior landscape consists of giant organs, cavernous bones, endless vistas of tissue, and geometric shapes that appear seemingly at random. Etched onto every sinuous surface are words such as *desire* and *cynicism,* while the tissue itself is stretched, striated, twisted, and eviscerated. *Landscape, environment,* and even *space* are terms that cannot articulate the peculiar experience of being surrounded by fragments of a body's interior that do not necessarily add up to a whole, and of discovering that this "inside" is already abstracted through text. From a different point of view (pov), these fragments appear enclosed in transparent cubes hooked up to various medical instruments.[32] For the VR voyager this pov clarifies the peculiarity of wandering-amongst-organs: the gigantic and grotesque body parts are now "framed" by medical science—the organs now represent an organon.

However, as the voyager twists, Möbius-like, from a position of engulfment and confusion to a view from above, the pleasure of resolution and containment is suspended, while the security of a subject position neatly distanced from the uncanniness of gigantic organs is denied.[33] For here, neither interiority nor abstraction are stable: the view from above depends on a wayward flick of the dataglove more than on any overarching narrative strategy, while the "ground" that interiority signifies sign is interrupted by inscription. Carved first by medical science and then by language, the interior body in this virtual environment, even in its grotesquery, is thus covered in culture. Yet the knowledge or resolution that culture suggests is not legible, chiefly because of the difficulty in controlling navigational speed or direction. Ironically, the voyager's own untrained body prohibits a reading of this other body as a coherent body rather than a collection of inchoate parts. Even when the transparent cubes represent a corporeality contained through science, what is seen is no longer a body but the detritus of dissection. And the voyager is painfully aware that any transcendence offered by a particular pov is entirely arbitrary:[34] an accidental flick of the wrist or a conspiracy of circumstance ultimately determines the subject position the voyager is able to occupy.

This accidental intervention of corporeality as it determines virtual existence is reminiscent of Cronenberg's *The Fly.*[35] In a postmodern interpretation of the original version[36] (where human arrogance and deception perpetrate the fatal combination of human body and insect head), Cronenberg centers the human/insect exchange at the genetic level—what one might call the *ur*-text of the body. He shows the genetic mutant as the result of accident, circumstance, a peculiar freak of nature whose very possibility arises from the computation of genetic code—"flesh made data." Part of the terror of *The Fly* stems from the grotesque proportions of the accident: proportions that humanism cannot reckon with, let alone contain or frame. Similarly, in Gromala and Sharir's *Virtual Bodies,* the arbitrariness of the pov points, metacritically, to the arbitrariness and hollowness of any notion of mastery, motivation, inherent source, or frame of reference that the body presents. Not only is narrative resolution denied but, more importantly perhaps, the arbitrary event, like the accident, refuses any kind of dialogue. Unlike the phantoms of early telephony or the memory constructs of Gibsonian cyberspace, the accident occurs in the absence of a clear connection. As a gigantic organ, a decaying Brundle fly, or even a planetary heritage, the accident—despite its abstraction—is mute.

In this space of disconnection, the accident also returns us to pain. But there is a sense in *Virtual Bodies* that pain, like sound, is *subject* to its articulation; thus pain is intertwined with different forms and structures of the grotesque, the absurd, the unspeakable. One cannot feel another's pain, and pain itself can never be adequately represented. However, this lack is repeated here at the level of effect rather than affect, as a cynical enunciation of an act of torture that takes the form of a stutter—a kind of biodigital glitch—rather than a scream or cry. Perhaps this is why Sharir's dance, both in the VR piece and in the presentation of the work as performance, was so compelling. His minimal movements showed a dancer's body constrained by familiarity with the VR gear, while the peculiar motions of his hands—derived from the gestures of "American death artists" with whom he has been working for the past fifteen years—seemed to be grasping, as the virtual hand grasps its object, but in a way that was both distracted and compulsive. This did not surprise me, since my first thought in watching Sharir's dance was of an image—either seen or dreamt—of an obese man supervising the work of the death camps, obsessively

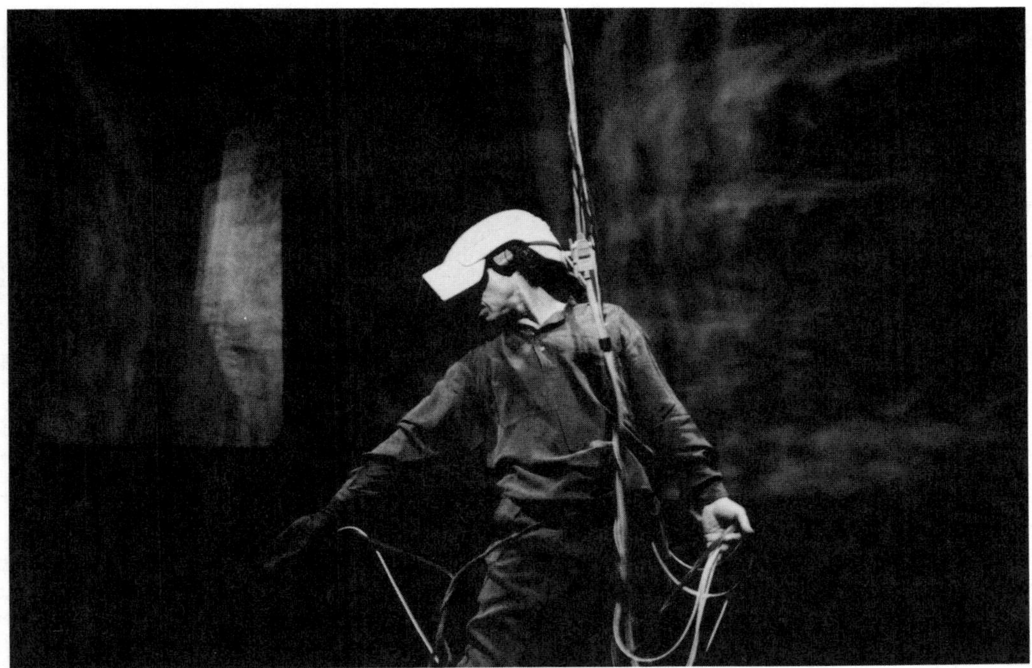

16. Diane J. Gromala and Yacov Sharir, *Dancing with the Virtual Dervish: Virtual Bodies* (1994), performance detail with Yacov Sharir in foreground and virtual worlds projected in background. Photo Donald Lee.

rubbing his thumb and forefinger as he sat at a desk, impatient for the work to be done, for the death camps to have finished their gruesome mandate, for the inevitable retribution that would be his own termination.

The compulsion to tap out the end of existence was a central theme in Antonin Artaud's *There Is No More Firmament,* itself an adaptation of the composer Edgar Varèse's unrealized project, *L'Astronomie.* Set in the year 2000, the narrative centers around the end of the earth initiated by a scientist who willingly annihilates space through "celestial telegraphy" in order to establish "interplanetary language."[37] Varèse's initial vision was to create through sound a feeling "akin to that aroused by a beam of light sent forth by a powerful searchlight."[38] The figures of telegraphy and "beams of sound"[39] evoke a sense of the inscription of space, which, for Artaud, means the end of the universe. Beams also appear throughout Marcos Novak's component of *Dancing with the Virtual Dervish: Worlds in Progress* and, like Varèse's projections, also incorporate sound.

At first glance Novak's *Worlds in Progress* seems to be the antithesis of Gromala and Sharir's rendition. Instead of an inscribed, contained, and dissected interiority, Novak's virtual environment presents an absolute exteriority: the infinite blackness of space dissected by lines of light—pulsars—and inhabited by mathematical constructs. Flying through this space is a dizzying experience, as constructs, worlds, or chambers speed by with the pace of a circuit city advertisement. Indeed, the work could be read as a sophisticated interactive rendition of the visual tropology that is currently used to represent cyberspace. Abstract forms, zooming trajectories, illuminated lines against a backdrop of outer space give the impression that one has entered a predictable, rational, and knowable universe. Here is certainty, grounded in the a priori axioms of physics and commodity culture. Here is a space where flesh could become data without any grotesque residue, where the subject could be "reduced to bits, represented in the system and in the process become information anew,"[40] thereby authentically entering cyberspace.

However, whether by accident or design, such a familiar dream does not materialize. Once in a chamber, I found myself surrounded not by the reassuring ambiance of mathematics, but by an unnerving sequence of auditory and visual pulsation that tugged at my field of vision—that is, my world—as if the color, light, and sound they emitted were gelatinous. In particular the sound, designed to interact with the head and hand movements, interrupted any vision of abstract space, as its low frequencies and relentless pulsing rhythm reverberated within. By accident (I am told), the design for audio interactivity configured the voyager upside down so that when one pointed one's hand and head upward, expecting pitch and rhythm to rise, the reverse happened.

Perhaps I was experiencing the "pulsars" that Novak referred to in his presentation of the work as lines that "create a kind of heartbeat" controlling other events.[41] Perhaps this was an example of what Novak calls the "intimate infinite"—the sense of "being there" that shields the voyager from the blackness or nothingness one encounters when first putting the helmet on or looking in the wrong direction. Perhaps it was an example of the "intonation of space,"[42] a liquid architecture, to use Novak's term, that approaches music, or a music that approaches architecture as sonic beams projected into space. Heartbeat, pulse—what is this if not the return of the body, palpable as sound, to smear the fine lines of exterior, objective, and knowable space with the confusions of the interior? What is this cloying mix of sight and sound if not the return of the ectoplasmic, phantom, and grotesquely hybrid forms disconnecting the smooth human/machine interface, interrupting the official discourse and flailing outside the computational grid? Does the infinite rendered intimate now become another form of the gigantic, presenting an undifferentiated space that confers not omniscience but the absence of markers, the nonownership of territory? Is this not a space of abjection—a space more aural than visual, more palpable that knowable?

If only by accident, the pulse opens a portal to compulsion, which Novak implies existed from the inception of the work. As the artist explains, on one hand, "Like a bird of prey, my acuity allows me to glide high above the planes of information";[43] on the other, "I feel the weight of histories . . . and this piece is intended to immerse itself in that."[44] For Novak, the clear lines of light that build these new spaces are themselves contaminated: light is now weighty, the abstract now grave, and the dense matter of the flesh now enmeshes the

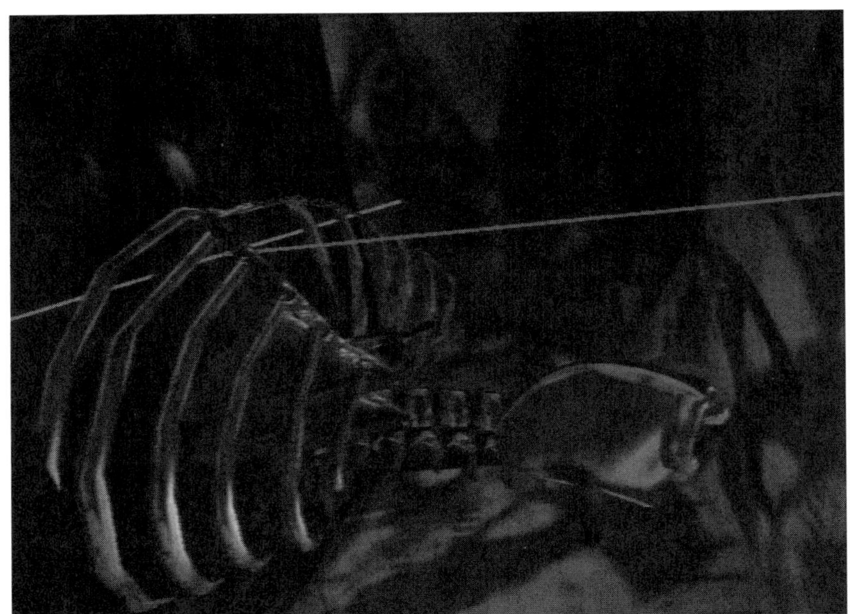

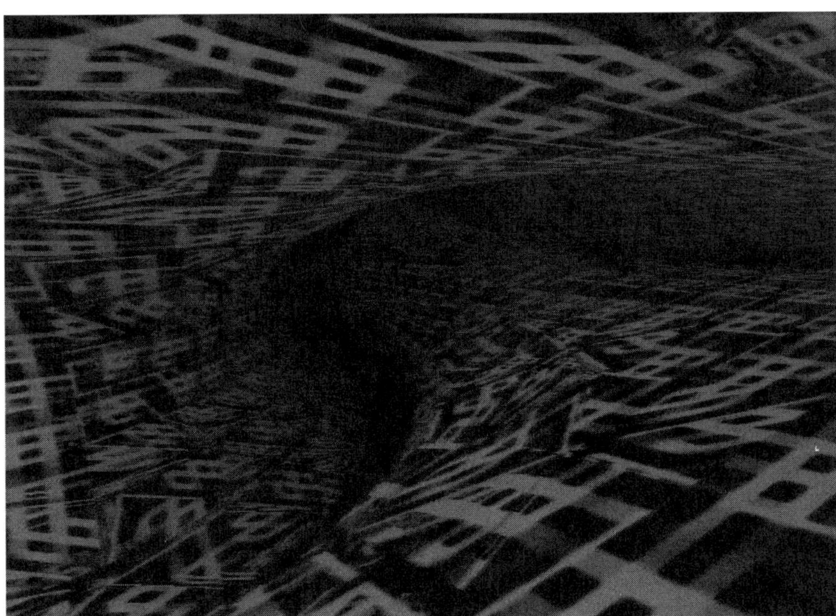

17. Marcos Novak, *Dancing with the Virtual Dervish: Worlds in Progress* (1994), frame grabs of virtual worlds. These show two of the chambers. Digital images courtesy the artist.

voyager in ethics. Novak concludes his presentation of the work by pointing out a floating window that screens images of "the smiles and eyes one sees in advertisements, Sufi patterns, and images of tortured people from the files of the Gestapo in Athens."[45] He interprets this final inclusion as "bringing back the body to the disembodied," but adds somewhat enigmatically, "I don't know what to make of this except to say that as far out as I go with these virtual things, that cruelty that we're capable of keeps nagging and I had to put it in."[46]

I also had a double dose of nagging. The beams of light and sound populating the universe, the dense inscription of the body, the randomness of frames of reference and points of view, the slippage between accident, anteriority, the grotesque, and the posthuman, the movement from the mute to the mutation—all seemed too suggestive of the strange hand gesture repeated in Sharir's dance and in my dream. Thinking again of Artaud—whose body literally convulsed with electroshock treatment as it was symbolically pierced by language—I realize that the telegraphic tapping out of the end of the world in an obsessive bid for the absolute frame of science is an extreme way, not of ensuring death, but of finalizing or terminating the *time between pulses*. That "between" represents all that is accidental, unforeseen, all that denies the futurity of vision and calls instead for a perspective from behind. To terminate the tapping would be to secure a fully inscribed, fully knowable universe where there is literally *no space* for either the grotesque or the uncertain.

As Elaine Scarry writes, there is no space for uncertainty in pain,[47] but in this context pain is itself riddled with nagging doubt. What pulls and tugs on the periphery of the frame, be it the computation of flesh or the limits of existence, is the absurdity of pain's articulation, and the cruelty that understands such absurdity. Within this matrix, the "lines of light" notating the virtual universe and engineering the posthuman subject appear to be striated with the desire to secure language, articulation, and inscription as the end of corporeality. These lines are no longer purely diaphanous like the table rappings associated with the spiritist era of telephony: they signal and they sound. But further, as *Dancing with the Virtual Dervish* reveals only too well, in the architectonics of posthumanism lines are becoming pulses—pulses that are incisive, that pierce the intellect as they have already pierced the ear.

NOTES

1. I use *Dasein* here in its Heideggerian sense to refer to sound's here and now presence of being-in-the-world.

2. See Christian Metz, "Aural Objects," *Yale French Studies* 60 (1980).

3. In his introduction to *Cyberspace: First Steps,* Michael Benedikt characterizes the "wireless," a forerunner of virtual reality, as a medium "not unlike the air itself—surrounding, permeating, cycling, invisible, without memory or the demand of it, conciliating otherwise disparate and perhaps antagonistic individuals and regional cultures." Michael Benedikt, ed., *Cyberspace: First Steps* (Cambridge: MIT Press, 1991), 10.

4. The Reverend L. Osler of Providence, Rhode Island, preached that God had reserved the discovery of practical uses of steam and electricity to "those closing days of this world's history, when the King's business would require haste"; cited in Carolyn Marvin, *When Old Technologies Were New* (Oxford: Oxford University Press, 1988), 126.

5. The change in meaning of "broadcast" at the onset of the communications revolution inaugurated by the First World War is discussed in Tom Lewis, *Empire of the Air,* PBS documentary, September 1991, KAET; and Benjamin Woolley, *Virtual Worlds: A Journey in Hype and Hyperreality* (Cambridge: Blackwell, 1992).

6. For more on the connections between telephony and Judeo-Christain mythology see my "Circuits of the Voice," *MusicWorks* 53 (1992).

7. According to Daniel Czitrom, the relation between the wireless and ether stirred anew the old dream of universal communication, "a dream expressed in religious terms by the early commentator on the telegraph. But with the wireless, the discussion drew its metaphors and vocabulary from physics and biology." Daniel Czitrom, *Media and the American Mind* (Chapel Hill: University of North Carolina Press, 1982), 65.

8. Walter Ong, *Orality and Literacy* (New York: Methuen, 1982).

9. For Bachelard, "Radio uniquely presents the possibility for communication between the human psyche . . . the unconscious . . . [and] the foundations of human originality. . . . [Radio] is not merely a function transmitting truths or news. It must have an autonomous life of its own in this logosphere, this universe of speech, this cosmic conversation which is a new reality for man. . . . Radio must find a way of bringing 'unconscious's' into communication. It is through them we will find a certain universality." Gaston Bachelard, *The Right to Dream* (New York: Orion, 1971), 188–190.

10. Ibid.

11. Being both unseen and at home are the key factors in the awakening of a kind of interior archetype of "homeness" that, not surprisingly, complemented the commercial construction of the domestic sphere within which radio was now "part of the furniture."

12. Rudolf Arnheim, *Radio* (New York: Da Capo Press, 1972), 76.

13. Ibid., 78, emphasis mine.

14. *Der Prix Ars Electronica* catalog (Linz: Veritas-Verlag, 1992), 104.

15. For example, a "typical" sound shape is sketched for an instrument, then only its "attack" (generally occurring in the initial part of the sound shape's curve) is sampled and repeated or "looped." The theory behind this practice is that the attack is what makes the instrument recognizable *as* a violin, or a tuba, for example.

16. Brenda Laurel notes that "even though we don't see sound, virtual reality researchers commonly speak as though we do. For example, sound is 'displayed' with audio displays. We don't know where this tendency to use visual metaphors for aural experiences started, but it was long before virtual reality." Brenda Laurel, *Computers as Theatre* (Reading, Massachusetts: Addison-Wesley, 1993), 125. In the discourse, sound is represented as *signal,* a term that connotes movement, only in terms of its input and output from the computer onto some other storage device, while in the computer it is referred to as *data.* With the sound wave represented imagistically, there is a tendency for viewing to become dominant over listening, the eye over the ear. While both ear and eye

are already inscribed organs, the potentially radicalizing effects of aurality, both as a physiological, resonant phenomenon and an ontological, cultural paradox, suffocate under the homogeneity of what could be called "oculo-logocentrism."

17. As Rick Bidlack describes, "Typically, these sound sources are attached to graphical objects in the virtual space; as the user navigates through the space, the location of these objects, and hence the direction of the virtual sound sources, change relative to the user's point of view. At a higher level, an interface has been designed which allows the creator of the virtual space to specify the behavior of specific sound objects in the space, as well as to create a responsive and dynamically changing aesthetic sonic environment as the user interacts with the space." Rick Bidlack, Dorota Blaszczak, and Gary Kendall, "An Implementation of a 3D Binaural Audio System within an Integrated Virtual Reality Environment," unpublished paper, Banff Centre for the Arts, 1993.

18. As Randolph Begault writes: "You could describe the sound of a reverberate environment by showing a direct sound and its 'early reflections' (the first bounces of the sound off the walls) coming towards you. Then you calculate the early reflections' angle of incidence to a listener and you apply head-related transfer functions to the delayed sound as well as to the direct sound. Then we simulate dense reverberation, which basically is many, many reflections, by using exponentially decaying white noise with different distributions, and the result is spatial reverberation." Durand R. Begault, "The Virtual Reality of 3-D Sound," in *Cyberarts,* ed. Linda Jacobson (San Francisco: Miller Freeman, 1992), 83.

19. Ibid.

20. Scott Foster describes the process in reference to the Convolvotron, a three-dimensional audio system that he helped develop for NASA-Ames, and which is used primarily for virtual reality systems: "When you wear the headphones and peer into the stereoscopic display in a Convolvotron set-up, the system's headposition tracker lets you look around and move through a simulated environment, with the sounds adjusting accordingly. We can adjust the sounds' reflections and even the positions of the simulated walls at any time through the computer system, based on an IBM PC/AT. The head-related transfer function essentially represents a measurement of the transfer characteristics of sounds

from points in space to the eardrums of a listener. This measurement takes into account the effects on the sound of the head, pinna, shoulders, nose, paunch, and other body parts." Scott Forster, "The Convolvotron Concept," in *Cyberarts,* ed. Linda Jacobson (San Francisco: Miller Freeman, 1992), 93.

21. Begault, "The Virtual Reality of 3-D Sound," 84–85.

22. For an elaboration on the "inner voice" of western metaphysics, see my article "The Genealogy of the Radio Voice," in *Radio Rethink: Art, Sound and Transmission,* ed. Daina Augaitis and Dan Lander (Banff: Walter Phillips Gallery, 1994).

23. See Steve Aukstankalnis and David Blatner, *Silicon Mirage: The Art and Science of Virtual Reality* (Berkeley: Peachpit Press, 1992), 28, 102.

24. Laurel, *Computers as Theatre,* 207.

25. John Belton, "Technology and Aesthetics of Film Sound," in *Film Sound: Theory and Practice,* ed. Elizabeth Weiss and John Belton (New York: Columbia University Press, 1985), 66.

26. Linda Jacobson, "Sound Enters the Third Dimension," in *Cyberarts,* 78.

27. Jaron Lanier, "Life in the Data Cloud," *Mondo 2000* 2 (Summer 1990), 50.

28. Aukstankalnis and Blatner, *Silicon Mirage,* 120.

29. As Slavoj Žižek writes: "The ultimate lesson of the 'Virtual reality' is the virtualization of the very 'true' reality: by the mirage of 'Virtual reality,' the 'true' reality itself is posited as a semblance of itself, as a pure symbolic construct." Slavoj Žižek, "From Virtual Reality to the Virtualization of Reality," in *On Justifying the Hypothetical Nature of Art and the Non-identicality within the Object World* (Cologne: Robert Fleck, 1992), 135.

30. Friedrich Kittler, "Gramophone, Film, Typewriter," *October* 41 (Summer 1987), 106, 115.

31. Presentation of *Dancing with the Virtual Dervish I,* Art and Virtual Environments Symposium, Banff Centre for the Arts, May 24, 1994.

32. This latter articulation extends the practice of exhibiting body parts in glass jars—a practice that, as Susan Stewart points out, "eliminates the possibility of contagion, indeed of lived experience, at the same time as it maximizes the possibilities of transcendent vision." See Susan Stewart, *On Longing: Narratives of the Miniature, the Gigantic, the Souvenir, the Collection* (Baltimore: Johns Hopkins University Press, 1984), 68.

33. Stewart notes that "the gigantic transforms the body into miniature, especially pointing to the body's 'toylike' and 'insignificant' aspects." Ibid., 71.

34. As Katherine Hayles writes: "Cyberspace represents a quantum leap forward into the technological construction of vision. Instead of an embodied consciousness looking through the window at a scene, consciousness moves through the screen to becomes the pov, leaving behind the body as an unoccupied shell. In cyberspace, point of view does not emanate from the character, rather the pov literally is the character. . . . Reduced to a point, the pov is abstracted into a purely temporal entity with no spatial extension; metaphorized into an interactive space, the dataspace is narrativized by the pov's movement through it." N. Katherine Hayles, "Virtual Bodies and Flickering Signifiers," *October* 66 (Fall 1993), 83–84.

35. Hayles mentions *The Fly* as a point of transition from the human to the posthuman that also marks the arrival of "flickering signification," which "brings together language with a psychodynamics based on the symbolic moment when the human confronts the post-human." Ibid., 79.

36. Kurt Neumann's original version was produced in 1958.

37. The full quote reads: "STUPENDOUS DISCOVERY. SKY PHYSICALLY ABOLISHED. EARTH ONLY A MINUTE AWAY FROM SIRIUS. NO MORE FIRMANENT. CELESTIAL TELEGRAPHY BORN. INTERPLANETARY LANGUAGE ESTABLISHED." Antonin Artaud, "There Is No More Firmanent," *Antonin Artaud, Collected Works,* vol. 2, trans. Victor Corti (London: Calder and Boyars, 1971), 85.

38. Edgar Varèse, "The Liberation of Sound," in *Contemporary Composers on Contemporary Music,* ed. Elliott Schwartz and Barney Childs (New York: Holt, Rinehart and Winston, 1967), 197.

39. Ibid.

40. Marcos Novak, """Liquid Architecture," in *Cyberspace: First Steps,* ed. Michael Benedikt (Cambridge: MIT Press, 1991), 225.

41. Marcos Novak, presentation of *Dancing with the Virtual Dervish: Worlds in Progress,* Art and Virtual Environments Symposium, Banff Centre for the Arts, May 24, 1994.

42. Ibid.

43. Novak, "Liquid Architecture," 230.

44. Novak, presentation of *Dancing with the Virtual Dervish: Worlds in Progress.*

45. Ibid.

46. Ibid.

47. Elaine Scarry, *The Body in Pain* (New York: Oxford University Press, 1985).

Cyberdämmerung at Wellspring Systems

Allucquère Rosanne Stone

Fade in to:

A modern office complex complete with indoor atrium featuring a twenty-foot waterfall. Inside nestle the corporate headquarters of Wellspring Systems. The company's generous office windows overlook a nature conservation area not far from the company's door, complete with dense, deep green foliage and brilliantly plumed exotic birds. Wellspring is an extremely successful computer game company, one of the first to produce a widely popular game. It has followed up on this with a succession of popular programs based more or less on the same theme: shoot-'em-up games filled with intense action, high-tech war machinery, and conventionally beautiful women whose purpose is to be sexually available to the protagonist's gaze. In the programmers' lair, which at Wellspring is a long, claustrophobic hallway lined with narrow doors, life goes on as it has at game companies since the palmy days at the Atari Game Division, back in Silicon Valley in the mid-eighties. But there are notable differences. Gone is the freewheeling, anything-goes atmosphere, gone the enormous sense of expectancy. Not that the work isn't exciting. The doors in the programmers' lair open on the same tiny cubicles, and the tokens of the technokid culture are still in evidence—here some crushed bags of Cheetos and a stack of empty Jolt cans, there a dirty, rumpled sleeping bag unrolled underneath a long table stacked with untidy piles of software. The differences are not immediately obvious, but include such basics as the distribution of wealth. The programmers' situation at Wellspring is far removed from that of the Atari programmers, who were pulling down sixty to eighty thousand dollars a year on royalties in the

1980s when eighty thousand was a fair amount of money. Instead, the Wellspring kids work on fixed salaries, not royalties, and in the back of their eyes lurks fear—fear of getting old without having hit the jackpot with a runaway selling program: that is, fear of turning thirty. And fear of the hundreds of technokids waiting outside for one of the elect to stumble, to falter, to make that tiny fatal error that magically turns them into one of the outsiders and lets another outsider in. Such fatal errors include objecting to the killingly long hours and low pay.

The owners of the digital sweatshop grimly enforce the pressure-cooker atmosphere of competition. They are multimillionaires. Sam Lerner drives to work in a Lamborghini Countach, a $150,000 ultra-high-performance sports car that gets ten miles to the gallon. He is thirty-two years old.

Unlike the golden days of Silicon Valley, there isn't even a pretense of trickle-down economics in the new world of software games. When Wellspring was acquired by a major, well-established game company with main offices in Silicon Valley, there was a celebration at the offices, with cheap cake and soft drinks. During the ensuing congratulatory speeches, one of the programmers asked whether some of the enormous profits made by Lerner during the sale would be distributed in the form of bonuses. Lerner looked at him as if he were some new species of vermin and said, "Of course not."

Nine years after the women at the Atari Lab first asked why there weren't more women in game programming, one hundred percent of the programmers at Wellspring are male. There is a single woman associated with one of the development teams, though not as a programmer. Susan Macintosh is in her early forties, a dynamic, fiery dialogue writer who aspires to be a producer, which is the person who actually determines the shape and purpose of a game. Over the years she has written a series of intriguing game plots designed to interest players of both sexes in something more complex than shooting at enemy blips. She presents these as possible games, and the programmers routinely ignore them.

During a recent development session Macintosh finally convinced one of the staff producers to take on such a project. She'd been trying for three years. After the meeting she left on vacation for a week, and returned to find that the project had been scrapped. "Too complicated," her boss said. "We replaced it with a shoot-the-helicopter game."

Macintosh patiently points out that the helicopter game has no plot. Of course she's patient. If she weren't patient, if in fact she weren't possessed of the patience of a certified saint, three years of this would have sent her screaming wacko.

"That's okay," the producer says. "People don't want plot. They just want to shoot things."

Memphis Smith is a short man of medium build. A bit older than most of his charges on the programming team, he is in his early thirties; unlike them, he is highly fashion-conscious in a Texan way, given to silk shirts, blousy pants, and silver-toed cowboy boots. He has shoulder-length jet black hair and a short black beard. Smith is a supervisor, in charge of an entire game production team. He has been with Wellspring from the very beginning, way back in the dark ages, in 1985. The game he is supervising is called Battle Commander, which promises to be the biggest seller in Wellspring's history.

During Wellspring's most recent new product rollout at Comdex in Las Vegas, the company debuted a pre-release version of Battle Commander. Comdex was Smith's finest hour. As the game's supervisor, he'd spent months honchoing his producer and pack of programmers along. He'd tracked and herded every facet of its development, and was now at the Wellspring booth, decked out in his customary silk shirt, receiving the plaudits of the throng. Nearby was his wife, keeping just out of the way. The brilliant arc lights in the vast vaulted ceiling glittered in the polished silver toes of his black Justin boots.

The opening screen for Battle Commander shows a naked young woman covered by a thin sheet, lying on an army cot. When the cursor passes over her, she sits up and looks seductively out at the player. Smith was particularly proud of this screen. At Comdex he was usually surrounded by a knot of chatting, admiring young men, and he hardly noticed that the few women who stopped to watch the demo shook their heads in disappointment or disgust and walked away.

The second morning of the Comdex show a woman from the marketing division of Electronic Arts went up to Smith and asked politely, "How might I go about influencing the way women are depicted in this game?"

Smith looked her over as if she were a putrefying fish. He inflated his chest just a bit—something of an accomplishment, considering his already cocky attitude. "Well, little lady," he said in an exaggerated drawl, "tell you what, why don't you just take it up with the artist, or better yet"—he leaned in at her, pushing his face close to hers, his voice dripping sarcasm—"Why don't you just call my boss and get me fired?"

There was a pause. "I see," she said, and walked away.

" 'The way women are depicted in the game,' " Smith chuckles. "You can always tell the ones that never get any."

His wife nods in assent. She is perfectly coiffed; she wears full makeup even when swimming. She is wearing the sort of dress that a wisecracking 1940s detective might have said was sprayed on. In her flawless mask, styled blond hair, high heels, and improbably form-fitting costume she bears a startling resemblance to Mattel's Barbie. In response to Smith's witticism she smiles, a bright, perfect, dazzling, crimson-lined Barbie smile. She agrees with him completely. It is a moment out of "The Stepford Wives."

A few weeks after Comdex, Sam Lerner's phone rings at Wellspring headquarters. He grabs the receiver and finds himself on the line with Larry Pierce, president of the company that has just acquired a controlling share of Wellspring stock from him. Pierce is curt, and wastes no time on introductions.

"Change it," he says.

"Change what?" Lerner asks, puzzled.

"The loading screen on Battle Commander. Change it."

Lerner, who has had things one hundred percent his way since he started the company, says blandly, "I really think that's our decision to make, Larry. We designed the game, and we know the market."

"Change it."

Smith comes roaring into Bomber Area, fists balled, bootheels pounding the carpet. The first person he sees is Chris Roberts, the brother of Battle Commander's producer. Roberts is clearly startled to see Smith so enraged. He strides up to Roberts and snarls, "Larry says to change the opening screen!" Smith looks wildly around the bullpen, shoulder-length black hair whipping at his face. He reaches up and pushes it back, the veins in his arm bulging around the gold band of his Rolex. "Somebody is trying to polarize this team. I know it's them damn frog women."

Frog women is Smith's term for women who do not meet his standard of attractiveness, which includes any woman with the temerity to remark on the quality of his games. "Those damn frog women don't like what we're doing. Stirring up trouble like that. Where do they come off saying Battle Commander is sexist?" He pounds his fist against the door. "There is absolutely no sexism in this game!"

He begins to pace, the silver points of his boots kicking wads of paper out of the way. "Never. Never in hell. Never will we change this game for a bunch of flaming assholes!"

Roberts calms him down. They decide on what they believe is a compromise. The naked woman stays. A second cot appears in the screen, on which lies a naked man, also covered with a sheet. He does not sit up when cursored.

Electronic Arts was one of the few game companies to evince a serious interest in what makes games salable to target groups other than adolescent boys. Trip Hawkins and Larry Probst, two of its principals, have given talks here and there on what their research has shown. "We don't want to change the world," Hawkins says, "but we feel we can do a better job and still increase our market share. We've been doing some research into gender differences. This stuff is still in the beginning stage, but I can report on a little of it. For instance, we know you can get a boy to go out and kill and destroy (in a game) for the sake of his Commander, or the President, or the country. Or maybe just because he feels like killing. In other words, boys will kill for the sake of an idea. But we also know that you can get a girl to go out and kill (in a game) to rescue a wounded child or a parent. In other words, boys kill for the sake of ideas, and girls kill for the sake of individuals."

EA is researching a line of games for young women. And while the amount of publicity this activity has generated may be greater than the results have warranted, it is a beginning. Other, more exciting research by other organizations is unfortunately still secret as this article goes to press. Recently, however, another game software company seized on this information to begin researching a line of mindless, bang-bang-shoot-'em-up games for girls.

Fade out.

A friend and colleague was looking over my shoulder as I wrote the Wellspring piece. "This story put me in mind," she said, "of an underground cartoon character of the early 1960s, a black superhero named Captain Melanin. Melanin was in Georgia, helping out with the sit-ins. When the police chief closed in to arrest him, Melanin said a magic word at him: 'Boomlay Boom.'

"The chief stopped, brainfried. Melanin said suggestively, 'We're all going to live in harmony now, right?' And the chief said, 'Hell no, I've decided to arrest myself instead.'

"Melanin looked out of the frame at the reader, shrugged his shoulders, and said, 'Oh, well . . . little by little.'"

Cut to:

Oshkosh, Wisconsin, Municipal Court. People have been standing in line since before dawn to assure themselves of seats for the trial of a man accused of raping a woman with multiple personalities by coaxing out one of her compliant alters. A few had brought folding chairs to use while they waited in the predawn chill. Some had huddled amiably on beach blankets with thermos jugs of steaming coffee. The composition of the crowd is extraordinarily diverse.

Finally the courtroom doors are opened and the crowd files inside. After an agonizing wait while people chat with each other with the same lively animation I associate with waiting for the start of a long-anticipated film, the bailiff calls the room to order. The silence is instantaneous. "All rise," the bailiff calls, and Judge Hawley strides in, followed by the court stenographer.

Hawley sits down in the high-backed leather chair, squares something on his desk, and looks down from the bench at the packed courtroom, his glasses catching the light. The sound of people getting seated dies away, and a hush again falls over the room.

For the most part Hawley has not said very much beyond what was required of him as presiding magistrate, but this morning he clears his throat and makes a brief introductory speech. His voice carries well in the room. It is a calm voice, not too inflected.

"Before we proceed any further, I want to make sure all the video and film equipment in this room is turned off and that all the cameras are down out of sight." He scans the room slowly, more for effect than for surveillance, then continues in the same calm voice. "There has been an unusual amount of attention surrounding this case. The issues we are considering are of an unusual nature. But I want to make it clear to everyone here that this is not a circus. This is a very sensitive case. There may be some bizarre behavior that you have not witnessed before. But nothing should get in the way of this being a court of law, first and foremost. I know that I can expect you to behave appropriately."

Nods from the spectators. People settle deeper into their seats. The unusually large population of professionals among the spectators now makes itself known as people reach into bags and briefcases for their yellow notepads, making the room bloom like a gray field dotted with buttercups.

Hawley nods to the prosecuter, Joseph Paulus. The silence deepens, if that were possible, and Paulus calls his first witness of the day.

Sarah walks briskly to the stand. She seats herself and is sworn in. She puts her hands in her lap and looks calmly at Paulus. This is the main event, I think. It is what this whole thing is about, really. It is not about columns in a newspaper. It is not about theory or discussion. It is not about sound bite media hype. It is about a young, calm, slightly Asian-looking woman in a short-sleeved white cotton sweater and a calf-length pale blue skirt.

Paulus stands a few feet in front of her, holding his body relaxed and still. He speaks to her in a normal conversational tone, not very loud but clearly audible in the silent room.

"Sarah, you've heard some testimony here about some events that took place recently in Shiner Park. Do you recall that testimony?"

Sarah nods slightly, then adds, "I do."

"Do you have any personal knowledge as to the events in the park?"

"No," Sarah says, "I do not." Her voice is quiet, flat, matter-of-fact.

"Who would be in the best position to talk about the events in the park that night?"

"Franny," Sarah says.

"Would it be possible for us to, uh—" Paulus hesitates and looks like he wants to clear his throat, but he settles for an instant's pause instead and then continues—"meet Franny, and talk to her?"

"Yes," Sarah says, looking calmly at him. A beat or two. "Now?"

"Yes," Paulus says. "Take your time."

The silence is absolute. Faintly, from somewhere outside in the hallway, something metallic drops to the floor and rolls.

Sarah closes her eyes and slowly lowers her head until her chin is resting on her chest. She sits that way, her body still, breathing slowly and shallowly. It seems as though everyone in the room holds a collective breath. The muted hush of the air conditioning comes slowly up from the background as if someone had turned up a volume control.

Maybe five seconds passes, maybe ten. It feels like hours. Then she raises her head, and slowly opens her eyes.

She looks at Paulus, and suddenly her face is animated, alive and mobile in a way that it hadn't been a moment ago. The muscles around her mouth and eyes seem to work differently, to be somehow more robust. She looks him up and down, taking him in with obvious appreciation. "Hel-lo," she says.

"Franny?" Paulus says, inquisitively.

"Good morning," Franny says. She looks around at the windowless courtroom. "Or good afternoon—which is it?" Her phrasing is more musical than it had been, with an odd lilt to the words. It, too, is animated, but it doesn't sound quite like an animated voice should have sounded. Also, on closer inspection it appears that the more animated look of her features hasn't spread down into her body. Her posture, the way she holds herself, the positions of her shoulders and legs and the relative tension in the muscles of her body, hasn't changed very much from Sarah's posture.

Paulus looks as if he wants to feel relieved, but again he hides it quickly. "It's, uh, morning, actually," he says, in a conversational tone. "How are you today?"

"I'm fine. How are you?" The same lilt to the words.

"Just fine. Now I was just talking to Sarah a few moments ago, and I'd like to talk to you about what happened June ninth of 1990." He glances up at Hawley. "But before we do that, the judge has to talk to you."

Hawley looks down at Franny. When she faces forward most of what he could see of her is the top of her head, but she turns now to face him. Her expression is hard to catch, but Hawley looks perfectly placid, as if swearing in several people in one body were something he does every day. "Franny," he says, "I'd like you to raise your right hand for me, please."

Hawley swears her in, his face impassive. It sounds like any other court ritual.

Fade out.

Stay tuned for an important message from your theoretician.

Multiple personality (MP, without the stigmatizing final D) is a mode that resonates throughout the virtual communities—of which cyberspace communities are part—that have been my field sites since 1981. For reasons that I find not entirely clear, the foremost researchers in MP dismiss out of hand the idea that there can be more than one person in a single body. Instead they fall back on received social and cultural norms concerning the meaning of "person" and "body." Like the surgeons at the Stanford Gender Dysphoria Project, of whom I have written elsewhere, MP researchers act wittingly or unwittingly as gatekeepers for meaning within a larger cultural frame.

In this context of multiplicity and psychology, transgression and constraint, it is useful to consider the work of Sherry Turkle. In her study "Constructions and Reconstructions of the Self in Virtual Reality," presented at the Third International Conference on Cyberspace, Turkle notes:

> The power of the [virtual] medium as a material for the projection of both conscious and unconscious aspects of the self suggests an analogy between multiple-user domains (MUDs) and psychotherapeutic milieus. . . . MUDs are a context for constructions and reconstructions of identity; they are also a context for reflecting on old notions of identity itself. Through contemporary psychoanalytic theory which stresses the decentered subject and through the

> fragmented selves presented by patients (and most dramatically the increasing numbers of patients who present with multiple personality) psychology confronts the ways in which any unitary notion of identity is problematic and illusory. What is the self when it functions as a society? What is the self when it divides its labor among its constituent "alters" or "avatars"? Those burdened by post-traumatic dissociative syndrome (MPD) suffer the question; inhabitants of MUDs play with it.

In Turkle's context of virtual systems, the question that many MP researchers dismisses as, to them, obviously false—namely, whether multiple selves can inhabit a single body—is irrelevant. The socioepistemic structures by means of which the meanings of the terms "self" and "body" are produced operate differently in virtual space. Turkle not only perceives this acutely, but seizes upon it and turns it into a psychotherapeutic tool. Moreover, she shows how the uses of virtual space as an adjunct to therapy translate across domains, beyond the virtual worlds and into the biological. What in this context might be called the ultimate experiment—plugging an MP into a Multiple-User Object-Oriented Domain (MOO)—has yet to be performed. Thus we have not yet observed one of its possibly hopeful outcomes: healing trauma but preserving multiplicity, or, perhaps more pertinent, creating discursive space for a possibly transformative legitimation of some forms of multiplicity. The answers to the questions Turkle poses above, as well as to those suggested by the brief congeries of narratives I provide here and elsewhere—such as, why MPD is so important to an examination of communication technology, and whether there is room for nontraumatic multiplicity in clinical accounts—are bound up in fine with the prosthetic character of virtuality. The technosocial space of virtual systems, with its irruptive ludic quality, its potential for experimentation and emergence, is a domain of nontraumatic multiplicity. Turkle and others, myself included, are waiting to observe how the dialogue between nontraumatic multiplicity and clinical accounts emerges in a still-to-be-defined therapeutic context.

The questions raised in Oshkosh, which Judge Hawley described as splitting some very fine hairs, were serious attempts to open space for new interpretations of what counts as a person in a world in which fragmentation and multiplicity occasionally push themselves into public awareness. Indirectly, the Oshkosh event points out the troubled locus of conflicting forces that marks out the space of the individual in modern society. How these forces merge and separate, and the battles between them, are perspicuous engagements in the war of desire and technology—the troubling and productive emergence not of new social but new personal formations, emerging from the contested zone at the boundaries between the machinic assemblages of political power and the inexpungable human drive for sociality and love.

There is also a lot at stake for us in the Wellspring story, with its grim real-world look at what contemporary game programming is like. It points out that there is no mandate in our culture to do anything in particular with the powerful technologies we have at our command. Of course information technology in all its wonderful forms is one of those. The computer game industry suffers from a feedback loop no more and no less pernicious than any other in a market-driven economy, which is that it's very easy and low-risk to go on endlessly making games for the same market. Leaving aside for the moment the problems of continually reifying xenophobia, the overwhelming majority of those games denigrate women, deliberately or offhandedly. Quite a few people, including people of both of the major genders, have tried for quite a long time to bring about even modest change in that regard. They haven't been very successful. In large part (though not entirely) this is due to the character and habits of the people who actually program the games. In our research at the Advanced Communication Technologies Laboratory at the University of Texas at Austin, my students and colleagues find that the programmers, almost exclusively young adolescent or postadolescent males, tend to live their lives in the same manner in which they write their games—with singleminded determination and a very narrow set of goals. Usually they have little in the way of social lives; they don't read books, but occasionally read comics; and they tend to perpetuate extraordinarily immature ideas of personal interaction styles. One of these is the way they relate to women. Women tend to be the same kinds of objects for them in their lives as they are in their games, and this is the heart of the problem of how pernicious the loop is: they don't believe there is a problem, because it's invisible to them. Under questioning, they intensely resist acquiring insight. They believe there is no sexism in their games, just as they believe there is no sexism in their lives.

This situation isn't surprising, really, although something in my heart keeps telling me it should be. The reality is that in the past fifteen years or so the computer game industry has gone with explosive speed from a few boys making up games for their friends to a few more boys making up games for millions of other boys. The level of sophistication hasn't changed, but the scale has changed drastically. Ethics haven't kept up, in fact have never been an issue since the thing is market-driven and kids as a market are a relatively new phenomenon. An unanticipated product of this meaning-machine is a particular kind of monistic identity, and the contradiction here should not pass unnoticed: one of the most powerful tools for opening identity into richly textured diversity winds up being a constraining force. If this reminds us of the evolution of television from the 1950s, when programming was still experimental, on to its maturity as a medium for selling goods, it means we are paying attention.

This loops back to the question of the ludic dimension of human-computer interaction, which I've discussed at length elsewhere in other essays. Should things like computer games, which are so terrifically absorbing and which take up so much waking time—so much precious, irreplaceable waking time—be expected to possess a modicum of invention, to be able to stretch players' imaginations and skills beyond the ability to hit targets and dodge obstacles? (Not that those aren't valuable skills, but they aren't the only valuable skills either.) Should we expect play to be edifying, or on the other hand will kids manage to make anything that comes to hand edifying against all odds? Is the game market simply a dead loss from the standpoint of sexism and education, something we have to learn to live with like we had to learn to live with the Bomb? How is it that the very young, the very talented, don't perceive the incredible power for change that has fallen to them by default—and the hideous consequences of failing to grasp that weapon when it's offered? How very like the hero mythoid itself, the symbol-cluster of the Quest that underlies so many heavily gendered stories of conquest and victory.

The critical importance for our purposes lies in the simple aphorism that software produces subjects. When humans engage with a symbolic structure of sufficient complexity, to a certain extent they synchronize or harmonize their own internal symbology with that structure. In this we are merely carrying out our own programs as social beings. Software is merely a perspicuous example; as McLuhan pointed out seemingly so long ago, any medium serves perfectly well in this capacity. The result of being immersed in such a medium is a gradual synchronization of a certain rigor of symbology with the symbolic structure of the medium and, for lack of a better term, a synchronization of voice. Some of you will recall that before the signing of the Nuclear Non-Proliferation Treaty, the Bulletin of the Atomic Scientists had a clock on the cover. For many years the clock was set very close to midnight—to Doomsday. Sometimes I think that destroying ourselves in a nuclear holocaust may be a better way to go than some of the other possibilities that go less remarked. The quiet death that comes when we are completely spoken, when others speak us rather than ourselves being able to speak, is for me the most frightening. That's why for me prosthetic communication and the things it creates, specifically interactive entertainment software, the Internet, cyberspace, and virtual reality, are not a question of market share or even of content. In a fundamental McLuhanesque sense these things are parts of ourselves. As with all powerful discourses, their very existence shapes us. Since in a deep sense they are languages, it's hard to see what they do, because what they do is to structure seeing. They act on the systems—social, cultural, neurological—by which we make meaning. Their implicit messages change us.

That, in fine, is how a few crazies at keyboards wind up influencing the social behavior of millions of game players, and why we end up splitting finer and finer hairs in order to keep opening spaces for more breadth and fluidity in what counts as a person. Is the answer, then, to outlaw violence and sexism in games? Almost certainly not. The band-aid approach of prohibition has a lousy and ineffectual history. For example in the United States, since the Comics Code, which prohibits extreme violence in comic books, became the de facto industry standard in the late 1950s, violence among young persons has increased several hundred percent. If there is, in fact, some way to remold society closer to the heart's desire, it shows no visible signs of having anything to do with curing bad behavior by not looking at all those "dangerous" images. It's not the images, and it's not the technologies either. Neither, simply by itself, is going to "save" society. In fact, the idea that technology in and of itself is going to change the world for the better, as the myths surrounding VR would have it, is merely pernicious. It's just reifying technology as holy grail, turning salvation into hardware, something condensed and made visible through the tellings and retellings of the cultural myths particular to us. These myths are extremely powerful, because they carry in veiled form the entire cultural force and imagined progressus of our western societies. But, as the knights who searched for the Grail found out the hard way, if it's salvation that you simply must have, you will never find it in a physical object.

Thus some of our culture's most enduring storytellers tried to tell us, in their beautiful and twisted way, that the end of oppression, racism, sexism, or, for that matter, binary opposition as a concept, is no more to be found in virtual worlds than it was to be found in computers, electricity, mesmerism, sex, drugs, or rock 'n' roll. This suggestion merely points more directly to the question of what or who gets to make technology whatever it is, whether it's okay or not. We do that. Save sheer accident, technology has no force outside of a system of social practices. If the world is to be changed, we are going to have to do it—for better or worse, with our eyes open, in the full knowledge that we ourselves are either destroying or creating our own future. No technologies, no illusions, no masks.

I look forward eagerly to continuing this high adventure with you—the adventure of our own future, as we immerse ourselves ever more deeply in our own technologies, as the boundaries between our technologies and ourselves continue to dissolve, as we inexorably become creatures that we cannot even now imagine. It is a moment that simultaneously holds immense threat and immense promise. I don't want to lose sight of either, because we need to guide ourselves—remember cyber means steer—in all our assembled forms and multiple selves right between the two towers of promise and danger, of desire and technology. In the space between them lies the path to our adventure at the dawn of the virtual age, the adventure that belongs to our time and that is ours alone.

Afterword

This paper was written under interesting circumstances, and I want to specially thank Mary Anne Moser at Banff and Bob Prior at MIT Press for their forbearance and simultaneously their gentle encouragement during that time.

A Disappearance of Community

Avital Ronell

This paper concerns virtual reality, media, and the war in the Gulf. I may not master the materials, or even the immaterials (VR opens the question of immateriality). Yet this issue is about mastery and I consider myself an infomaniaque. VR is philosophically complex even though it is system-dependent on classical tropes of representation, imagination, the sovereign subject, and negated otherness (negated otherness is what Hegel called the enemy). So it is dependent on a number of metaphysical cravings, but then who isn't? Might as well face it: there has always been a desire to transcend the body, and I have often said that I would donate my own body to science fiction. But the donation of a body in life is part of a metaphysical striving toward an indeterminate elsewhere. VR is jamming the master codes of this historical desire for transcendence and exteriority, and deserves to be heard out in all its effusions. Because VR is also about being-in-the-world and liberating the location of being to nonsubstantial spaces, it is trying to reconfigure the possibilities for sharing the world. According to Jaron Lanier, Howard Rheingold, and others, VR practices a politics of finitude[1] and demands an ethics of technology. At the same, however, VR is jacking into Mattel Corporation, the military, and NASA but it is *also* a design testing ground for architecture and the medical sciences. "So, where were we?" asks the scholar. A rigorous study of virtual reality cannot be made to fit entirely into the frame of this paper, but I will try to raise some questions to indicate the direction of my reflections on this difficult topic.

Virtual reality, artificial reality, dataspace, or cyberspace are inscriptions of a desire whose principal symptom can be seen as the absence of community.

Lanier's discourse has everything to do with possibilities for constituting an ecstatic community that would refuse to be governed by a goal or a totalitarian drive toward unification and project. It is as if Lanier were tempted to retrieve Bataille's community of shattered egos. To a certain extent, the metaphysical subject is broken up and displaced to routes of splintering disidentification. The subject no longer finds its mooring in identification with a substantial image. Still, Lanier, like all transmitters on behalf of radio (VR is said to be more like a telephone or radio, disconnected from the television apparatus), emphasizes the ego-building prowess of VR designs. This is a double-edged claim that we need to probe with unrelenting clarity.

In cyburbia there is always a risk of blurring the distinction between simulation and the operational world. I am not convinced, though others are, that this entails a very new risk. More seriously, perhaps, there is still a tendency to retrofit the technological prosthesis to a metaphysical subject—the sovereign subject of history, destiny's copilot. In other words, the technological prosthesis would merely be an amplifier and intensifier borrowed by a centered subject whose fragmentation is, as they say, a simulation—that is, a device for *disavowing* fragmentation, selflessness, or, on another register, castration. This aspect of disavowal in part explains why Lanier, for his part, would have preferred to see VR called, as he asserts, "intentional reality." Now this exposes a whole problematics of the intentional consciousness that I will address momentarily. The intentional consciousness is a philosophical construct that emerges when it is felt that we are not in control of our actions. And this is precisely where Lanier's project takes hold, in the control rooms of his majesty the ego. In addition to arguing that VR is "good for the ego," he launches an assault against the contemporary subject's passivity; this assault on passivity—quite understandable at first sight—is however made in the dubious name of action and control.[2] The subject will take control. The argument against passivity, which I can only graze swiftly, is, I believe, a false and highly problematic one, though I also agree with the need to activate radical creativity. But the opposition of passive and active proffers a deluded equation. Take a look around you. Have we not, as a culture, been too active, too action-filled, even if action splits itself into representations of the traumatized spectator and manic warrior? Any uninterrogated invention made in the name of action has got to attract our deepest suspicions. A true ethics of community, whether located in cyberspace or among lovers, readers, artists, activists, and so on, would have to locate a passivity beyond passivity—a space for repose and reflection, a space that would let the other come. Exposing oneself to the other, or to the other's death, has nothing to do with action as such. I will elaborate this shortly. It is because of an articulated desire for action that cyberspace is also west of the west, that is, a Memorex cowboy frontier.

One of the errors that Lanier makes is to say that "ultimately, everything is done by people and technology is only a little game that we play."[3] The war has shown us that people do not play little games. Nor is technology zoned outside of us, but, as cybernetics and artificial intelligence have shown, man-machinic hybridizations involve a refashioning of the human being, following a trajectory of altogether new alignments of self and the technobody. Now, the body, from Marvin Minsky onward, has been devalorized into the "bloody mess of organic matter." I am not an essential feminist, but I do think that this utterance could never have been made by a woman, who takes out monthly mortgages on her body in the form of periodic bloodbaths and PMS. This is being said too quickly, perhaps, but it is urgent to recognize that the body of a woman has a fundamental relationship to death, despair, finitude—and to life. While the woman's body produces the eternal return of the "bloody mess of organic matter," the cyborg soldier, located in command and control systems, exercises on the fields of denial. Intentional reality eliminates the body as organic, finite, damageable, eviscerable, castratable, crushable entity, thereby closing the orifices and stemming leakage and excrement. We are not very far from Deleuze and Guattari's BWO: the body-without-organs. Orificial shutdown and excremental control help explain why the Gulf War was conducted under the compulsive sign of cleanliness: on the American side of language usage, this was a clean war, a clean-up job accomplished according to the moral, political, and military evaluations that were represented. It was so clean, in fact, that there were blank screens assuring the protocols of propriety—the covering up of coverage—but also it was so clean on our side of the line in the sand that the American body, if it was to be lost, did so in an ascension of friendly fire. The point is that the other side never got to us, which I shall try to analyze briefly as the immunopathological dimension of this war.

For the duration of the war, contact with the negated other was nanominimalist and the language of contact was suspended. Thus, even the inevitable contaminations implied by linkage were avoided. No linkage means, among other received things, no parasitical or random eruptions on the mainline of firing, but also no complexity and no ambivalence concerning this war, its aims, or the aims of man (the new order is about the aims of man, and we know which way they are aiming). The disassociation of communicating parts—no linkage—reflects the effects of derealization that the war continues to produce on us.

So where were we? Before I say more, I want to say that I am assuming that we are still in North America, which is to say that I take it for granted that your opinions, evaluations, and feelings concerning the Gulf War have been formed. If at this point I say that we are still in North America, this is in part because the war in the Gulf has destabilized our understanding of location, and has instituted a teletopical logic: a logic of spaces aligned according to technological mappings, where the newer is far and vice versa, and where systems of boundaries and borderlines will have to be entirely rethought.

If we are still in North America, which is not only a place but also a time of reflection—a site where geopolitical and chronopolitical tensions are being played out—this also means that we are particularly sensitized to economies of justice that the war has rendered transparent. In this area of discourse which we share, certain economies have come to light, but it is the light of apparent contradiction. I shall take one crucial example.

We constitute a community of readers and speakers who have stressed time and again that the cost of war has drained the resources of AIDS research, or at least we have noticed that our friend with AIDS are not receiving the support they urgently require. We have wondered collectively and singly about the displacement of funds from health concerns to the demands of the military. It appears that there exists a contradiction between the external and internal needs of the United States. However, this is not a contradiction, nor a blind spot, nor even a fundamental displacement of a particular psychic investment. The lack of AIDS support and the war investment are part of the same experience of a national desire to the extent that the war was guided by a rhetoric of renewal and regeneration, in other words, the war was conducted entirely within the symbolic registers of fascistic health. In his essay entitled "Our History," French philosopher Jean-Luc Nancy has argued that an "ideology must be called 'fascist' in the general sense in which themes of spiritual and national regeneration, of the vigorous recovery of health through firmness and discipline, correspond to a fascist or fascistic vision of things."[4] What this means basically is that in the name of symbolic health, a unity of world that sees its image in wholesomeness and the project of renewal, we have waged war on what was repeatedly represented as degenerate, sickly, something that carried the threat of contagion. In this regard, America has been carrying out its newly transcendentalized project of killing the unwell, the contaminated. The enemy is imagined as being disorderly, inefficient, tactically illiterate, dysfunctional; and to a certain degree the projected solution, cybernetics, promises to overcome such instabilities.

The hygienic project has everything to do with establishing a new world order that consists in nothing less than purifying an imaginary—and real—territory, guaranteeing that it be proper, that it espouse values of propriety and property. The invisibility of the enemy inhabits this logic, which is essentially viral. Our war body has not only tested itself, but it has come out of this test site relatively intact, clean, healthy. This accounts for the other apparent contradiction—our insensible casualty rate, their massacre, our surgical strike, their bodies, our high-tech shoots, their blood. This was a translation, on a world-historical scale, of the AIDS test on which we scored HIV negative, because this was a safe war, run by the teleologic of what can provisionally be called "virtual reality"—which is almost an anagram of viral reality. Some of you may know that virtual comes from the Latin *virtus,* strength, manliness. Meanwhile, back at the Gulf, Americans have once again instituted an autoimmune laboratory. I wish to emphasize that I am equally analyzing the symbolic effects of the war according to the letter of the war, which is to say that I take the theater of operations, the rhetoric of surgery and other health metaphors, quite seriously. I even take seriously the fact that the Bush family had its own private theater of thyroid operations, which, as if his disease had externalized itself, is called in medical terms a thyroid storm. Commander-in-chief of one storm, the then-president is internally dominated by another storm. This in itself suggests how difficult it is to locate the origin of the storm—inside the deserted presidential body, in the resurrection of Nazi terminology (storm troopers), or in the projections onto the desert.

On a less unconscious level of corporeal transmission, the desert has been conflated with woman's body. This brings me to my next question. Why have we stormed the figure of a woman's body, why have we entered this mysterious legacy? Given the constraints of space, I can only point to where we might go in order to explore the imaginary contours of a feminine body that our forces stormed (of course you know that before they went up, fighter pilots were fed attack doses of pornography. This, together with drugs, would help them to drop on the feminized body of Iraq while at the same time we were protecting Kuwait from rape). If this were a seminar, I would ask you to turn your attention to Jacques Lacan's essay on "Aggressivity and Paranoia in Psychoanalysis," which is in *Ecrits*. Here you will find Lacan interpreting Melanie Klein's excellent work on the coordinates of original aggressivity. Through Melanie Klein we know the function of the imaginary primordial enclosure formed by the *imago* of the mother's body; through her we have the cartography of the mother's internal empire, the historical atlas of the intestinal divisions in which the *imagos* of the fathers and brothers ("real or virtual," says Lacan), in which the voracious aggression of the subject himself dispute their delirious (destructive) dominance over her sacred regions. What I would have to point out with regard to the particularity of the Gulf War, which may or may not be generalizable to other wars, is that the figure of the mother was always prominent, on both sides ("the mother of all battles" and so on), and that it was always understood that we were in a region of some originarity—the site of primordial aggression, the sacred origin of all culture. Because we were in a zone of primordial encounter, it was also read as a place of Armageddon and apocalyptic showdown. This was the end that was also to designate a beginning: the *new* world order. As the original war, it encompassed all the wars in modern history: the world wars, Vietnam, et cetera.

Because the war assumed this status as origin, the initiator of the new world order, and was being conducted in the womb and cradle of our civilization, it is not at all far-fetched to read with Lacan this war in its rapport to the subject's paranoiac mapping of the maternal body. A splinter of evidence in support of this view could be retrieved from the compulsive focus on the mothers who went to war. This insistence names the symptom—the mother's body—but in the mode of disavowal. Mother may have gone to war, but she was not the site of aggression; mother was finally on the map, but she was not the map itself or a conflictual site constituted by the imaginary. This leads us to ask more generally: What is the battlefield? What are its boundaries and symbolic localities. When does the battlefield take place? And how does it place us? What about the myth of the home front? And so on. In any case, Lacan in this difficult but crucial text establishes a link between space and aggressivity, which is to say that the domination of space is related to the narcissistic fear of damage to one's own body. In fact, he argues that the fear of death, the "absolute master" according to Hegel, is subordinate to the narcissistic fear of damage to one's own body. Aggressivity, as one of the intentional coordinates of the human ego, especially relative to the category of space (this includes real or virtual space), allows us to conceive of its role in modern neurosis. The preeminence of aggressivity in our civilization, Lacan adds, would be sufficiently demonstrated already by the fact that it is usually confused in "normal" morality with the virtue of strength. The glorification of strength as a social value is a sign of social devastation initiated on a planetary scale and justified by the image of a laissez-faire of the strongest predators. This condition should set off our psycho-alerts.

Now, the connectedness between virtual reality and the war was not always entirely evident to me. The promise of VR is immense, at times liberatory, and very careful in its articulated negotiations with metaphysics, technology, and play. I was initially perplexed because VR is something so new that it has only begun to display its existence; the war, on the other hand, seemed like something that ought to have been obsolesced, and in fact it was, though it did happen and it did tend to make claims for high-tech breakthroughs. This war incorporated many wars and was played out in a spectral battlefield: WWI (the gas masks), WWII (the calculated resurrection of Hitler), the Vietnam syndrome. These phantom wars that participated in the Gulf War even included the war we did not have with the Russians, or, for that matter, with the Martians. Can we live without an enemy? But if indeed there is something new about this war at a time of felt closure (most of us figured that the conditions for real war were vanishing, and that war games were a residual symptom of a history of battle), if we still feel that we are in a time of closure, then we have to recognize that closure is not the same thing as the end. Closure does not simply close a domain or an epoch: by tracing the limits of its possibilities, the closure also reaches the other side of its limit, exposing itself to its own exterior. In "Our History," Nancy has shown that there is no simple opposition of exteriority of the closure to the opening, and this is perhaps where VR comes into the picture.

A question that VR poses, in its full positivity, is where to locate the community. Because we are vanishing. In the absence of the polis, something like VR obligates us to pose ethical questions about contact, memory, the prosthetic subject, and it teaches us to dislocate our proper place.

There is no proper place: this includes ghettoes and kitchens, and all corresponding systems of the proper place. The politics of a room of one's own has to be rethought today, however enlightened it was yesterday. The question is a hard one, surpassing as it does the video game logic of good versus evil, winner and loser, presence and absence: can there be an atopicality of the community that nonetheless gathers, a community going nowhere, but ecstatic, a community of shattered egos, where the control towers come tumbling down, and where the other is genuinely anticipated? By this I also mean the other technologies.

NOTES

1. See Timothy Druckrey, "Revenge of the Nerds: An Interview with Jaron Lanier," *Afterimage* (May 1991). Some of the other utterances that invite interpretation include:

"This technology has the promise of transcending the body, depending on what you think a body is." "There's no object that could be less biological or messy. It doesn't have blood, it doesn't fart, it doesn't have eczema. The slick blackness of technology is a way of avoiding the messiness of the body. I think that's why sometimes you see more men associated with this technology than women. Men find themselves in a more desperate situation with the flight from death." "From a political point of view the clear precedent for virtual reality is the telephone." "I view VR as a fancy telephone in many ways." "The young generation who saw the political stakes in struggles over representation . . . now they have to face up to its indeterminacy. Is there something like an ethics of technology?" "What I'm hoping the virtual reality technology will do is sensitize people to these subjective or experiential aspects of life and help them notice what a marvelous, mystical thing it is to communicate with another person."

2. John Barlow, "Life in the Data Cloud: Scratching Your Eyes Back In" (interview with Jaron Lanier), *Mondo 2000* (Summer 1990), 2. "The computer is a map you can inhabit. Which is very seductive. It is mostly seductive because you love what you have to struggle for, but it is also seductive because it makes you seem very powerful. So it is good for the ego . . . a way for people to get ecstatic and be with each other." While the ego remains a problem, I do wish to signal that being-with (*Mitsein*), ecstatic temporality, and being-called are crucial issues in the philosophy of Martin Heidegger. Lanier is also inclined to plug the telephone. "The telephone is a total win." If Lanier is unwilling to count the losses, and keeps tallying technology according to egological scoreboards, this *may* be traceable to the fact that his mother was a victim of destructive technologies and suffered the dehumanizing effects of the Nazi camps. While I do not feel that it is necessarily appropriate in this case to psychoanalyze VR's unconscious genesis, I have tried to configure these relations—the maternal function, technology, and the terroristic state—in *The Telephone Book: Technology, Schizophrenia, Electric Speech* (Lincoln: University of Nebraska Press, 1989).

3. Ibid., 49. Also see Adam Heilbrun, "Virtual Reality: An Interview with Jaron Lanier," *Whole Earth Review* (Fall 1989).

4. Jean-Luc Nancy, "Our History," *Diacritics* (Fall 1990), 20.

Fascination, Masculinity, and Cyberspace

Rob Milthorp

This essay is about men and technology, particularly adventures in digital imaging and communication technologies and their relationships to the construction of masculinity in Euro-North American society. I write about the topic from the viewpoint of a white, heterosexual, middle-class visual artist who employs electronic media to stage visual and auditory environments as a way of exploring my own assumptions about being male. It will be evident that I am presenting a notion of technology specific to the historical and social practices of men and that I am not addressing the contributions and relationships of women to technology, nor exploring the diversity of gendered identities or the possibilities of alternate genders. These omissions have been made as part of a strategy to explore aspects of masculine gender roles. My definition of the term masculinity is inevitably reductive and specifically linked to men, although I recognize that masculine and feminine gender traits are present in varying degrees in members of both male and female sexes.

My use of the term cyberspace refers to a range of technological actualities and possibilities from high-tech virtual reality applications to banking machines and phone sex. It acknowledges the process of transformation from analogous experience to the organization of digitized information. More than that, however, I am working with a concept of the imagination, an abstract, tantalizing fantasy that has become part of a contemporary cultural reality. The relationships between fantasy and experience, expectations and fulfillment, science and our cultural mythos, are inextricable aspects of my notion of cyberspace.

Fascination

When Neil Armstrong made the great leap for mankind, I watched spellbound as the television images showed awkward cybernetic organisms shuffling through the moon dust. I was a young man who like many of my male friends had nourished my imagination with science fiction, from the classic works of H. G. Wells to the sexual fantasies of Edgar Rice Burroughs and the futurist tales of Isaac Asimov and Arthur C. Clarke. Any doubts I had about the apparent nature of the surrogate humans and landscape on my television screen was overcome by the thrill of the technological adventure of the moon walk; I was filled with satisfaction that the realms of my imagination were now much closer to reality. I felt pride and belonging, as if I were on the moon, that somehow I had contributed to this spectacular event, or that it had been produced specifically for me.

That uncritical and romantic experience was challenged several weeks later, when I visited my grandfather in England. Granddad was in his seventies, a spry and intelligent man who until his retirement had served as a guardsman on trains for British Rail. He was god-fearing and gentle, had been a nurse in World War I. Despite his connection with the great steam trains, which were also formative images of my boyhood, he lived for the most part without modern technology. Coal fires heated his house and toasted the bread. He did, however, have an ancient radio and an antique black and white television with which he kept track of the world via "Auntie," also known as the British Broadcasting Corporation.

I asked him about the lunar landing, anxious to hear his thoughts, anticipating the enormity of his excitement. His response was immediate and shook me. He did not believe the event had happened. He refused to watch the news clips anymore. He did not want to be convinced, with reason or passion. I was frustrated and angry, as if it were a personal affront. I was tempted to think that he was irrational, but I loved and respected him and could not make that judgment. It was a dilemma that turned into an important milestone for me.

I am aware today that my grandfather's bewildering attitude threatened beliefs that I took for granted but which I had not consciously identified, among them a psychological and material sense of identity with all things technological, which gave me security as a youth and an assurance about the future, as well as self-confidence as a man in the modern world. More subtle but equally in jeopardy was my sense of male camaraderie, of a common identity that transcended, and as I realize now, obliterated, the values of those who were not men of my cultural and racial heritage. If my grandmother had voiced the same disbelief as my grandfather I would not have been disturbed. However, the moon landing was a symbol that reaffirmed my personal desires and aspirations to succeed in the world I anticipated before me. Because my identity was so intimately connected to the beliefs that made the event possible, Granddad's unexpected response was incomprehensible and jarring.

Over time I began to understand that what I was willing to embrace as "actual" experience was not necessary nor acceptable for his reality. He had stepped outside the realm of my necessary beliefs, because in his final years we no longer shared the same aspirations. I began to develop seeds of doubt, began to question my own acceptance of the event and its relevance to my view of the world. Was technology as important to the world I desired as I had thought? I grew more sensitive to the distancing effect of technical, objective language and the authority that it carries upon men's personal communication.

Years later when I watched the *Challenger* shuttle explode I was again fascinated by the technological spectacle. I shed tears for the tragedy of the crew and was fixated by the death of Christa McAuliffe. Like millions of other television watchers, I was repelled by, and excitedly drawn back time and again to, the image of the explosion and the forked white vapor trail against the open blue sky. Questions about the marketing of the event, the necessity of the space program, the mechanisms by which the *Challenger* crew became as real as my own family stirred up uneasiness with my own association with the event. I felt uncomfortable, angry, torn between a form of sadness and a thrill generated by the violence of the event. I felt like a participant, but with an uneasy sense of responsibility, which was an emotional offshoot of the identity I had developed in connection with the whole notion of technological progress. This guilt was the flip side of the pride of ownership I felt for the moon landing.

I began to question the connection between my fascination with technology and what it meant to be a "male" male, that is, the imperative to know, to act, to be in control of events, to somehow reach toward objectives that could be formulated in terms that went beyond private or personal significance. Christa McAuliffe's death was at the heart of the disaster for me. She had not slipped into man's role as an astronaut, but remained a mother, a wife, a teacher. She transgressed the neglected border separating the male realm of technological adventure from the real world of people and feelings. Her role on that mission (it is hard to imagine it carried out by a man) was to teach science to schoolchildren from space via television.

Shortly after the explosion, and having worked as a visual artist for several years, I produced a videotape that included news clips of Christa McAuliffe walking to the spacecraft smiling and waving to the crowd, and tumbling in the womblike space of zero gravity in a high-altitude plane.[1] These and other images of the launch and explosion captured from television were juxtaposed with the image of a woman as she blew into a plastic toy pipe, suspending a ball above the bowl. The ball danced on the airstream, eventually slipping and falling to the ground. My intention was to parody and probe western society's obsession with technological progress, symbolized by the heroic spectacle of space exploration, and address the human costs. At the same time, more intuitively than consciously, I had identified men and masculine values with the attitudes toward technology that I was critiquing.

As an artist in 1994 these experiences are at the heart of my concern with issues of masculine gender construction, specifically the nature of men's communication and the integral relationship of technology and particularly communications tools with the characteristics of conventional masculine identity.

In my own work I use video, sound, and computer-generated images to investigate how male identity is defined by verbal and pictorial language. I work with the assumption that men have difficulty speaking openly or honestly about their feelings and desires, and that we consequently have difficulty in our relations with others—sometimes resorting to inappropriate expressions of control and power to achieve our objectives. My experience is that masculinity is preoccupied with doing, with affecting and controlling. I believe that actions for men frequently do speak louder than words. Technologies of countless types are tools for action, which serve men's objectives and achieve their needs but also mediate the direct communication that can be central to negotiation of mutual interests, to consensual sharing and community action. Men are often able to communicate most effectively when the communication is linked to task completion, or to competing for status—we are masters of how-to. However, we have greater problems with the affective communications that are part of responsible mutual relationships—the understanding of and empathy with what is—for ourselves as well as for others.[2] Obsession with technology can be both a symptom of and a license for men's avoidance of the vulnerability of unmediated communication. This is not to deny the benefits of technologies that sustain and enrich lives, but rather to question the cost of their use as a diversion from direct communication.

As a friend recently pointed out, modern instruments of communication are in part used as distancing devices, as a way of sublimating difficult issues of identity into a safer space, the location of secondary experience, in which knowledge and understanding are produced by discussing the evening television news with friends or colleagues at work the next day.

My own investigation of masculinity runs the risk of hiding itself within the novelty of electronic equipment and the confines of the gallery space. In 1989 I was surprised by a review of an installation in which I had employed video projections, programmed lights, and infrared switches to construct an interactive space for the viewer.[3] My intention was to explore intimate feelings and human relationships. The reviewer's conclusion was: "formally hot and emotionally cool."

The installation included a standing field of golden wheat that defined the space and the continuous projection of an aerial view of autumn fields. Two videotapes playing on monitors in the room presented the profiles of a man and a woman apparently in conversation, talking about their experiences in moving to a new home. Both spoke about their friends, those left behind and those newly encountered.

From time to time the profiles of the speakers turned from facing each other to facing the viewer. The movement of the viewer through the gallery activated the sound, so that at times the two would appear to be in conversation, their reminiscences moving in and out of phase, sometimes relating, sometimes not. Sometimes only one voice could be heard. The intention of the work was to address gender differences in experience and communication.

The overall visual aspect, as the reviewer stated, was forceful. The arrangement of the electronic devices worked well to accomplish my objective of moving the viewer through the space and, as I thought, to convey the experiences of Monte and Pam. However, the formal structure became the dominant feature of the work, naming the issues involved and in fact taking precedence over the affective relationship that was to be established between the viewer and the storytellers. As an artist I had taken the objective and safe role of director, pointing to but not identifying myself with the emotional substance of the work, because this was easier and less threatening, a way of avoiding the issues of my own masculinity by speaking for and about others.

While men's difficulty with communication is the concern underlying my research, I am also working with an opposing assumption, or perhaps hope, that most men share with me a profound discomfort with their role and responsibilities in relationships even though many of us have learned to deny this discomfort so completely that it is hard to imagine that we would ever want to change. While social power may be held predominantly by men as a class, and by individual men in specific circumstances, my experience is similar to that expressed by many men whose day-to-day experience of power is abstract, and who do not feel especially empowered but as often as not consider themselves to be subject to someone else's privilege within the social hierarchy. Paradoxically, it is frequently difficult for men who wish to address the balance of social power because they meet opposition both from the existing male world and from the critical positions that dismiss these unconventional male objectives as a sublimation of guilt or simply as increasing male power by giving it new voice.

The challenge for men in coming to grips with their position is aggravated by the images of men and male roles contained in the languages of contemporary popular culture, languages manufactured and distributed by communications technologies. Male violence, for instance, is or should be a major concern for men, considering that most violence is perpetrated by men against others. In the media, however, male violence is portrayed simply as violence, or youth violence or domestic violence.[4] The primary responsibility for violence is consequently deflected from the masculine to a generalized subject that all too often includes the victims. This deflection is made possible precisely by the reliance upon a mediated, impersonal technology with its attendant public codes and conventions to address the issues, while the actual violence that is being named continues without men holding themselves accountable.

These questions have led me to explore the obsessive nature of men's relationship with technology—a devotion that seems to alternately accommodate and then deflect the urgency of men's development of productive and mutual communication with others. The extremely sophisticated communications devices that are now being created could in theory assist men to develop better interpersonal skills for identifying, experiencing, and sharing with others. However, it is a tremendous expectation to place on the use of these tools to think that they can mediate entrenched psychological and social positions, particularly when the tools are themselves products of these values.

Masculinity and the Stuff of Technology

Like other men I am greatly attracted to the processes and products of technology, to "boy's toys." The late twentieth century has accelerated the manufacture of technological artifacts and our desire to own them. From cellular phones to digital recording, from portable PCs to Internet accounts, communications technology is the hottest consumer fashion. Cyberspace technology, romanticized by William Gibson and the holodeck of the U.S.S. *Enterprise,* is the latest passionate desire of the consumers of industrialized nations. Information is a staple consumer item of the late twentieth century. Interactive computer networks and digital technologies have shaped a community that relies on specialized languages and codes but to which access *seems* to have been democratized, at least among the dominant consumer classes. Ownership of the production and control of communications marketing represents power and inevitably male authority in an economy still dominated by men. But the attraction to technology is more complex than a desire for material power. Patriarchal social power is an aspect of the issues I am dealing with, though I will not directly address it here.

Psychoanalytic theory, and particularly object relations theory, has been helpful in attempting to explain the displacement of male affective relationships from one person to another, as the male child disidentifies from the mother, and counteridentifies with the father. When the father, responding to the socioeconomic constraints of capitalist society, is unable to sustain the relationship or be a spiritual guide or mentor, the son may transfer his passion from people to things; machines, technical processes, abstract ideas and symbols. People are objectified and devalued in his personal relationships. Relationships with other men are competitive and defined in terms of independence. Relationships with women and children are about power and control. Homophobia and misogyny are among the dysfunctional products of this inability to identify oneself with others.[5] From Carl Jung to Robert Bly to Nancy Chodorow a diverse group of writers and theorists acknowledge the impact on the psychological development of young men and women of the alienation of the father from the family in western society.

From our professional roles in the sciences, in research and development, in the abstract world of power and politics, to our passionate involvement with financial systems, sports, automobiles, and, more recently, the New Age realm of Cuisinarts and electronic kitchen wizardry, we are sustained by our relations with things—collecting, categorizing, knowing how things work. One example is the classic love affair some men have with Harley Davidson motorcycles, a symbolic identification not only because of the machine's power and outlaw status but because they can be lovingly cared for, decorated, known and controlled in all their functions. In an age in which information is power we are drawn to similar relationships with the tools of communication. Knowing a computer, all its characteristics, applications, foibles, and requirements, is equally fascinating for the hacker and for the novice.

Men's fascination with technology is linked to the masculine need to be in control of the material world, to know how to extend that control, to be able to act, and to be independent of reliance on others. In this respect science and technology are dominantly male in nature and have been forged from the development of capitalism, materialism, and individualism in western culture since the Renaissance. Reason is the supporting structure of the masculine fascination with technology and its products.[6]

Our understanding of civilization, progress, science, the separation and opposition of nature and culture, are dependent upon the philosophical ascendancy of reason in the seventeenth century. Technology was a means of objectifying nature. It served as an instrument of mapping, measuring, categorizing. It established viewpoints and determined quantifiable boundaries through devices such as Alberti's systemization of drawing using linear perspective. It articulated difference and imposed conformity. Reason, defined by men as a masculine domain, gave us power in the world and was the moral legitimator of the capitalist market system. This belief system was connected to a notion of the death of God and the ascendance of man as maker—ultimately as author and originator of knowledge and wisdom. Lack of reason, which would threaten the economic and political power structure, was equated with lack of humanity. Women, children, men who did not adhere to the prevailing notion of what was reasonable were seen as less than human and associated with the sensual world of nature.[7]

For individual men trying to live up to this definition of masculinity, the Cartesian opposition of emotion and reason, body and mind has meant a suppression of individual feelings, desires, and needs. Private experience that might subvert the hegemony of reason has been discounted, binding men to the institution(s) of men. As sociologist Victor Seidler writes, "The very identification of masculinity with reason has tended to blind men to their masculinity as something that has been socially and historically sustained."[8] The perpetuation of this condition depends on the society's success in sustaining a hegemonic system of power relations, which through its exploitation of human and other natural resources creates wealth and power.

The suppression of individual experience and emotions has been reinforced by the creation of public communities of men in such sites as science, the arts, technology, education, commerce, mass media, politics—with specific languages, structured to support reason, acting as the glue. The objectives of these communities, their functions, outweigh the private needs of the individual. The shaping of self-sustaining technologies by these communities results in languages that do not possess the words to accommodate the needs of difference. This is very clear in the example of women and medicine and the medical community's failure to address women's specific health needs. It is also present in the historical failure of the medical establishment to address men's health issues related to drug abuse, violence, suicide.

In these sites men have invented ways of communicating with each other to accomplish tasks and exert power that subordinate intimate experience and friendship to the abstract, to the universal, to the impersonal, but which at the same time appear to satisfy men's need to belong to a community.[9] Technologies are created to support these power structures.

As Daphne Spain describes in her study of men's friendships and power, men's huts in traditional societies allow elite men to congregate together, to compete amongst themselves, but also to provide a united front against the rest of the world. Their relationships with each other are not on private or intimate terms. The "technology" at work here is architectural, a real space in which the distribution of power is communicated. Women, on the other hand, tend to have fewer but more intimate relationships, which occupy private sites and do not have access to information or control of the technologies of power. Styles of communication mirror differences of gender priorities, as Deborah Tanner found in her study of sociolinguistic indicators of gender difference in communication:

> Intimacy is key in a world of connection where individuals negotiate complex networks of friendship, minimize differences, try to reach consensus, and avoid the appearance of superiority, which would highlight differences. In a world of status, *independence* is key, because a primary means of establishing status is to tell others what to do, and taking orders is a marker of low status. Though all humans need both intimacy and independence, women tend to focus on the first and men on the second.[10]

In industrial society, in which the imperatives of religion, food production, and procreation are no longer the simple or exclusive objects of power, men gather in their "huts" or institutional sites to establish power over material resources. Among these resources, technology and information control are key. In these communities men also exclude relationships or power-sharing with women or children or men who do not conform to the group. Tanner observes that this socialization pattern is built from childhood to maturity: "Boy's relationships are held together primarily by activities: doing things together, or talking about activities such as sports, or, later, politics. The forums in which men are most inclined to talk are those in which they feel the need to impress, in situations where their status is in question."[11]

As the institutions bind, so men seek refuge in them to reinforce their identities as men and to have support in sedating and keeping in check those feelings and emotions that clash with their definition of masculinity. It is a subterfuge, a "virtual" social space that has been created to uphold the objectives of power. John Grisham's popular novel *The Firm* presents a fictional and obviously stereotyped look at a cabal of lawyers who form such an institution.[12] The power of the leaders of the firm rests upon the suppression of the revolutionary power of the personal. Cyberspace may be the newest of these institutional sites, even though it appears to provide an anarchic space in which various kinds of hierarchies can exist and in which power is decentralized. Cyberspace may in fact reflect the increasing capacity of our culture to accommodate aberrant or marginal activity without destabilizing the existing structure. In a similar way student demonstrations and political actions of the 1960s and 1970s could be seen as effective means of containing social protest and potential anarchy, in effect contradicting their very objectives.

When my grandfather denied the lunar landing he was defying social control over his beliefs and values. He chose to read the docudrama event so that its significance fit the needs of his personal existence. He was resisting the greater community of consumers, in which we are all, men, women, and children, identified by our capacity to produce, to consume, and to be consumed, and in which information as knowledge is a key product. He was able to do this because, as a retired and relatively poor person, even though he was a man, he was on the margins of consumer society. Consumption defines and standardizes the terms on which we can relate—the class, gender, sexual, ethnic, cultural differences that determine who can belong, what language we can use, and what the price may be. My grandfather, being less important to the consumer system, was able to some extent to make an aberrant choice.

Among the costs of men's relationship with abstract systems and impersonal communications in institutionalized male social structures has been our separation from responsibility for our feelings as men. We suppress our sensitivity as weakness, and we place responsibility for our personal anger and aggressive behavior on others. Men are suspended between the compulsion to speak with their hearts (and risk being ostracized) and the institutional imperative to be silent and accepted (safe but inhibited). Communications technology, fashioned within male institutions, reflects this condition by offering the possibility of more free and open discourse but one still circumscribed by institutional control of form, language, access. Computer networks, cyberspace, and "virtual" technology, the latest and the best, heighten this contradiction because of the huge range of variables offered for the users' selection and manipulation. These increase the appearance of control and power for the user, because he can do more tasks, achieve quicker results, fashion working methods and styles closer to his tastes. However, the technological power, the control of the market and infrastructure, the source of the languages, while increasingly transparent, remain with the institution, not the consumer. In an analogous example, while more channels become available to the television consumer, they are still controlled by a handful of mass media conglomerates.

Authorship and Anonymity in Cyberspace

The theoretical social structures that bind men and technology are not enough to explain the sensuality, the aesthetic fascination that keeps men, whether "console jockeys," computer hackers, or tyros, in the embrace of their machines. Television has a weak attraction compared to the absorption that the computer holds for its devotees. Michael Heim identifies this as an erotic relationship in which technology exudes a "scent that once surrounded Wisdom." He points to a desire to emulate godliness, a condition of perfect knowledge.[13] Unable to attain that objective, men allow their passion and desire for pure information, the basis of technology, to substitute for the Platonic ideal of wisdom. The proliferation of network information banks on every subject imaginable, however specious, seems to satisfy the need to substitute knowledge for the kind of experience and understanding that is largely a product of interaction with others in the world.

The space of cyberspace is a construction, an architecture in which, seemingly, an operator can transfer its agency to a puppet, in a time and context governed by different rules than those of the "real" external world. It permits an interactivity that denies the body at the same time as it simulates it. Like the other public sites central to male identity, cyberspace creates the appearance of intimate knowledge and relations without the obligations and responsibilities of relations. It generates its own exclusive codes, languages, and behaviors. It represents escape from the world at the same time as it makes the world more visible and open to question.[14]

In my own surfing through both academic and popular news groups, the potential for men to constructively investigate the parameters of masculine identity does not appear to be realized, unless one counts the superficial exchange of names or video games featuring the possibility of assuming a one-dimensional caricature of another gender. As in fictional literature and film/video, it is more common for men to extend themselves as androids or cyborgs, as alternate identities, than it is for them to investigate personal objectives by shifting identity to different human identities. Robocop, Mr. Data, and the Borg in *Star Trek: The Next Generation,* HAL in Clarke's *2001, a Space Odyssey,* and Asimov's robot Daneel Olivaw investigate male roles from the viewpoint of artificial life forms, rarely from alternative human viewpoints despite the machines' aspirations to humanity. Men become machines and vice versa. The fascination still lies in an abstract technology. These are out-of-body experiences, symptomatic of men's tendencies to distance themselves from direct identification with others. As Robert Romanyshyn develops, a central aspect of the modern scientific world has been the transformation of the body into a machine separable from consciousness, mutable and ultimately subject to abandonment.[15]

One of the latest applications of virtual technology, no longer the sole territory of the military and NASA, has appeared in the financial sphere. Competing software and hardware researchers have developed virtual systems that allow market analysts to see and manipulate investments as "virtual" objects instead of as calculations.The analyst is faced with far too much information to handle; relying on the number crunching and organizational intelligence of the computer, he or she can make more intuitive decisions independent of the numbers themselves. The body extends the technology because of the body's limitations. The data is part of a larger system of information in which many participate, but the interface between individuals is in a singularly isolated realm. As the technology gets better, the achievements of the enterprise grow and the body becomes increasingly inadequate and dependent. The task-driven environment is not designed for exploration of self or others. For instance, in a flight simulator the amount of information to be processed requires that the machine monitor and make decisions without human intervention. The mind/body supervises but does not necessarily understand the actual process or significance of the operations, or the myriad of social operations that have combined to create this technology, its objectives and social significance. This is a frequent condition of complex scientific endeavors such as the development of nuclear weapons technology.

The lure and fascination of video game fantasy, already a multimillion-dollar business in the United States in 1993, is that of a vicarious escapist experience that also responds to men's technological passion.[16] Video games are largely based on conflict and win-lose competition. Multiple players can form relationships based upon the rules of play, but these relationships are limited by the same conventional "male" rules. The stockbroker, the Nintendo player, like the viewer of the ten o'clock news, really only develops meaningful relational experiences when returned to his body, to negotiate a new contract with his boss, to argue the choice of the next game with his brother, or to debate the effectiveness of NATO in Bosnia with his lover. It is in face-to-face communication that intimacy is achieved and that responsibility for mutual understanding is most likely to be taken. In the hierarchical world of status that is integral to men's social environment, distanced communication is familiar because it maintains independence. However, it deflects critique and self-reflexive awareness.

The computer/telephone information and communication network Internet was originally conceived by American defense experts to be decentralized, as a protection against the loss of communications and information in the event of nuclear devastation. The consequent but incidental democratization and freedom for users is in the process of being eroded as multinational corporations like AT&T, with its 3.2-billion-circuit kilometer telephone grid, and government agencies, particularly those concerned with crime and security, vie for control of the power and profits. Whatever opportunities the nets could provide for men to examine their social identities, by providing an unregulated and safe space for interactive communication in which to explore their understandings of others, may soon be made unattainable by the priorities of commercial ownership and consumer standards.

The most sophisticated virtual reality systems, like those envisioned by William Gibson, would similarly allow men to break the bonds of the body in a seamless world of synthetic experience, creating mythical worlds and experience. But what chance is there of that world being any less determined by power relations than the computer nets will be? The maintenance and development of communications structures of increasing sophistication will rely on large sources of transnational wealth and power. Control of freedom of expression, copyright, moral and behavioral standards will depend on the interests and tolerance of those sources. When breakfast cereal advertisements become part of the Internet, how will that sponsorship exert its interests in family values, in agricultural and environmental practices?

In the cyberspace of the foreseeable future the notion of originality and authorship, both of them aspects of masculine individuality and independence, is problematic. The boundaries of the cyberspace experience are limited by the quantity and quality of information that has been input. And what information or whose images will they be anyway? Even if there are opportunities for users to negotiate shared environments, the database will possess a history and vested interests.

However, originality and authorship are not only subject to the nature of the ownership of cyberspace technology, of the nets, virtual reality, or other applications.

In a culture where the proliferation of images has conflated primary experience and a representation, image-makers have begun to abandon their attachment to the blank page of imagination and creativity that was the heritage of nineteenth-century art. Both in life and in art, image-making in the late capitalist world is characterized by pastiche, construction and deconstruction, appropriation and alteration of existing images. The romantic notion of authorship is now mediated by the demands of a culture in which images, like information, are currency.

This might be a world in which novelty substitutes for a personal sense of wonder and the stimulation of encountering new experiences, where ultimately there are no surprises. This secondhand experience can give the illusion of being controllable precisely because it lacks the richness and unpredictability of life. For instance, the erotic relationship between men and the image of knowledge offered by technology, identified by Michael Heim, may be erotic because in the form of information and secondhand experience it is safe, expectable, under control, and so only a distant mirror of the eroticism of embodied experience, an artificiality made to appear real.

What characteristics will men's relationships in cyberspace take? Anthony Rotundo in his book *American Manhood* articulates four male character types that have developed in twentieth-century America as outlets for expressing newly legitimated male passion, as the gender conventions of the Victorian age gave way to a less affectively constrained definition of manhood.[17] Rotundo calls them the team player, the existential hero, the pleasure seeker, and the spiritual warrior. While each of these types accounts for an increased social acceptance of male personal expression, they are all still tied to a masculine individualism based on a hierarchy of competitive status and control. The team player turns his aggressive and selfish ambitions to cooperation with his colleagues. He is still the decision-maker, however. The pleasure seeker is the consumer at his pinnacle, sublimating masculine passion in possession. The spiritual warrior, who can be found in the mythopoeic literature of Robert Bly and others, is the man who reconnects with and masters his primitive ancestral passions. The existential hero, fighting the restraints and corruption of society, lives at the margin, a romantic, antiauthoritarian individual, imagining himself outside the power of institutions. All of these character types can be conceived in relation to the themes of technology and cyberspace. Although cyberspace applications appear to offer opportunities to form new collaborative directions, the technology may in fact simply reflect these variations in masculine control patterns. For the team player this may be represented by collaboration to achieve competitive status outcome in business, academic, or military contexts. Ownership is the essence for the pleasure seeker, who establishes material status by possessing the new currency of information. For the spiritual warrior, authorship and the possibility of creating one's own narrative give license to self-discovery. The existential hero fits well with the stereotype of eccentric, individualistic characters who design and populate cyberspace culture. In film, literature, and sometimes in real life, this character, whom I will refer to as the cyberspace hero, prevents wars started by incompetent military bureaucracies, steals from big corporations like Robin Hood, foments revolutions, becomes a warrior against aliens, introduces viruses into the establishment computer networks.

As an artist it has been tempting for me to discern a connecting thread between the character of Rotundo's "existential hero," the course of western modernist history and modern art, and the cyberspace hero in the world with digital technologies. As a male-dominated narrative, bounded by formal image-making technologies, populated by heroic individuals, the modernist quest is defined by the need for personal originality (hence individual power) and at the same time a desire for the security of control (authority over others). These objectives have been reflected in the culturally dominant art forms of the western world during the twentieth century, a key aspect of which has been an obsession with abstraction. Abstraction, as an approach to image-making whether spiritually or materialistically motivated, represented an arena of highly personal codes and languages that vied for supremacy in artistic practice, whether concerned with space, color, or design. Abstraction has long if not forever been a part of image-making; but it was the formulation of these techniques as cultlike systems, like cubism, that set apart modernist activity. Abstraction of images, like objectivity and science, reflects a systematization of beliefs and values that I would argue is an outgrowth of a masculine code of individualism, which reinforces patriarchal power and the myth of the importance of the individual man while in fact alienating that man from his subjective experience and his awareness of the importance of a truly collaborative identity with others.

The heroic male figures of modern art, such as Gauguin, Malevich, Pollock, Rothko, are part of a totalizing mythology of originality and genius. They could be described as existential heroes, individuals rebelling against society and constructing their own worlds. Ironically they are also figures frequently discussed in terms of their struggle with social and personal alienation. They were, despite their mythic roles, particularly subject to the stress of cultural determination. Jackson Pollock's tragedy was that he could neither achieve nor keep up with the identity of heroic rebel that was created and marketed for him, and was only able to escape beyond the margins of the societal institutions he was unsuccessfully rebelling against with his death, a final act of irrational defiance.

The cyberspace hero, like William Gibson's console cowboy in *Neuromancer,* plays a similar role.[18] In usenets, video games, virtual constructs, the nature of the freedom of action and imagination created may have more to do with the fantasy of male individualism than with the possibility of a critical repositioning of masculinity. The coding and ideological positioning of the technology is contained in its jargon, its literature, its mythology. In modern art dogma, the existence of new languages was enough to assert progress. The discussion around cyberspace often contains the same assumptions, concealing social meaning beneath the veneer of the new. The subversion of social boundaries that is implied in the literature of cyberspace is itself subject to virtuality.

Conclusions

The picture I am presenting strikes even me as being desperate, but I recognize the difficulties of the task of reconstructing masculine identity in a male world capable of endlessly reenvisioning maleness in its own image. The new technologies and potentials may not be problematic by themselves. Getting rid of the prevailing patriarchal values and ceding their power to alternate approaches is. Rather than reject or ignore technological imperatives, men need to find ways of developing their ability to relate to others, and to bring these understandings to bear in determining the form and purpose of the technologies. The values and objectives of those who do not share or benefit from the current scope of masculine power need to be included in shaping the future of technology. Effective *and* affective communication, which includes mutual understanding and recognition of gender-based differences in communication priorities, is important to the achievement of these goals. Cyberspace in any of its forms from text-based to image, and in its various degrees of interactivity, is not a disembodied technology. The body sits at the keyboard, wears the visor, develops the software. Communications technology is not a substitute for the intimate communication of the body, of the raw and holistic experience of the senses, and that improvement in intrapersonal and social relations is unlikely to be made purely as a result of technological advance.

Cyberspace is a new tool. It mirrors prevailing cultural values and is a powerful instrument in conserving a power structure of male values and masculine behavioral characteristics. This power is institutionalized in public sites of knowledge and endeavor, such as cyberspace technology, that create their own codes and conventions particularly geared toward status and hierarchical organization. Individual men have little chance of escaping the hegemonic control of male communities that provide both sanction and reward for conformity. Men's needs for independence and self-sufficiency are reinforced in these sites as long as the institutional function is not compromised.

In the twentieth century, awareness of men's self-denial and alienation from a healthy social identity has generated opportunities for men to reconsider their need to relate on a different basis. Communications tools, from radio to cybernetics, might assist this rethinking. However, there is a risk that we are creating tools that simply make the artificial appear more real, and that by their very nature disguise that artifice, permitting men to avoid coming to terms with others and their world in meaningful relationships. In effect new ways have been developed for masculine expressions of control and status to be realized, as heroic individuals or in hierarchically oriented collaborations. The technology and its products provide more and more objects of fascination and obsession that occupy men's lives.

Perhaps my grandfather was afraid that the developments in space exploration that I celebrated would threaten the masculine values that had originated in his day. My brave new world and his beliefs were out of step. Retired and at rest, he drew a line between control and change that he refused to cross. I wonder if my own doubts about the cascade of developments in cyberspace technology are a reluctance on my part to accept change. But just as the cobra fascinates its prey, holding it in thrall, in a balance of desire and fear until the victim escapes or is consumed, I feel compelled to learn as much as I can about this technology without losing touch with my desire to understand and enrich my experience as a man.

Notes

1. I want to thank Karen McLaughlin, Calgary artist and writer, for the discussion that clarified this point and helped me to understand why I made the tape *Time to Time* in the first place.

2. Deborah Tannen, *You Just Don't Understand* (New York: William Morrow, 1990), 24–25. Deborah Tannen points out the differences in communication styles and objectives between men and women, based on different sets of hierarchial environments men and women occupy. Power and accomplishment tend to underlie masculine communication, while friendship and networking structures underlie feminine communication.

3. The installation was entitled *Eclogue* and was part of an exhibition at the Nickel Art Museum, Calgary, entitled "Interior Presence," curated by Donna McAlear. Nancy Tousley was the reviewer for the *Calgary Herald.*

4. David Baxter, registered social worker, interviewed on Calgary CBC Radio, *The Eye Opener,* May 18, 1994.

5. Liam Hudson and Bernadine Jacot, *The Way Men Think: Intellect, Intimacy and the Erotic Imagination* (New Haven: Yale University Press, 1991), 49–53.

6. Evelyn Fox Keller, *Reflections on Gender and Science* (New Haven: Yale University Press, 1985).

7. Victor Seidler, *Rediscovering Masculinity: Reason, Language and Sexuality* (New York: Routledge, 1989), 14.

8. Ibid.

9. Daphne Spain, "The Spatial Foundations of Men's Friendships and Men's Power," in *Men's Friendships,* ed. Peter M. Nardi (London: Sage, 1992).

10. Tannen, *You Just Don't Understand,* 26.

11. Ibid., 85.

12. John Grisham, *The Firm* (New York: Dell, 1989).

13. Michael Heim, "Erotic Ontology of Cyberspace," in *Cyberspace: First Steps,* ed. Michael Benedikt (Cambridge: MIT Press, 1992), 69–80.

14. Robert D. Romanyshyn, *Technology as Symptom and Dream* (London: Routledge, 1989).

15. Ibid.

16. Richard Corliss, "Virtual, Man!," *Time* 142:18 (November 1993), 58–61.

17. Anthony E. Rotundo, *American Manhood: Transformations in Masculinity from the Revolution to the Modern Era* (New York: Basic Books, 1993), 284–293.

18. William Gibson, *The Neuromancer* (London: HarperCollins, 1984).

A City for Bachelors

Jeanne Randolph

Introduction

Technology is a map.
Technology is a grammar.
Technology is a blueprint.
Technology is a catechism.
Technology is a surgery.[1]

I often say that technology is a representation.[2] When I think about technology and its metaphors as an ideology, I name it the "Technological Ethos."[3]

> Really, the only freedom you must insist on is motion, the only desire, the desire for speed.
> Stephen Bingham, ex-president, Alias Research Inc.[4]

> A promising approach is to look at virtual communities as communications systems, the inhabitants as signallers and receivers. The costs and benefits of anonymity, identity deception, character stability, etc. can thus be discussed without relying on the vocabulary of the embodied world.
> Judith S. Donath, MIT Media Lab[5]

> When you've got the right Big Idea, then all that you think about and all that you do will align itself toward it, like magnetic particles, and you won't ever have to worry about doing anything irrelevant again.
>
> Brenda Laurel, from *Computers as Theater* [6]

If I were to psychoanalyze the Technological Ethos, I would talk about its symptoms. Traditionally, one applies psychoanalytic technique to a person, perhaps to a society,[7] never to a concept. In North America the psychoanalysis of a person or of a society usually implies that the symptoms are a manifestation of disease.[8] The "symptoms" here would be technology in general and virtual environments in particular.

Yet this is not the way I want to engage psychoanalytic theory with virtual environments.[9]

> If there is to be any validity in the search for the psychological dimension of an artwork, it must not be at the expense of the work's potential to change the conditions into which it has been received. . . . There is a possibility that psychoanalytic ideas can extend into a realm where they are used not to reduce artworks to a psychological derivative, but rather a realm where creative production suggests new forms for psychoanalytic ideas.[10]

From the outset, this suggestion that there could be new forms of psychoanalytic ideas must, nevertheless, always include a fundamental psychoanalytic belief—that the subconscious is real.[11] From the outset, my own research into the relevance of psychoanalytic theory to the relationship between technology and culture has almost exclusively been indebted to the object relations branch[12] of psychoanalytic theory, a choice that should not be taken for granted.[13]

What is so disturbing about the relationship between the Technological Ethos and contemporary media arts that psychoanalysis becomes relevant? What are the unique dimensions of art in the era of the computer that require psychoanalysis—and specifically object relations theory?

Object relations theory was invented in order to acknowledge that identity—the self, personhood—is impossible without interaction, literally between the infant and her/his mother;[14] identity or personhood forever depends upon interaction between our (lifelong) developing self and the phenomena that are available in our environment (people, objects, events, ideas, and so on). Personhood is impossible without interaction. We will not, according to object relations theory, just grow like Topsy.

Object relations theory was invented to acknowledge the extent to which survival in or by society is inextricable from the fabrication of justifications for survival. Inextricable from the fabrication of "meanings," beliefs, ideologies, religions, and so-called culture is the site of that fabrication.[15]

Psychoanalysis of the City

It is our most recent form of representational technology;
it is a symptom;
an instrument of a re-ordering of perception and of power relations;
it is a representation of human knowledge,
this new thing,
an awkward instance of the new writing;
it is the new and easy armchair way to have an experience;
it is yet one more effort to buttress the masculine ego
against its perceived foes,
the realization of the "techno-mythological" concepts
of the automaton;
a product of programming (writing),
which is a system of rules and regulations;
it is a fast way of going beyond one's limits,
something that has grown
out of a Judeo-Christian society,
a theatre that contains the ideas of all things
except, of course, the idea of itself;
not an ahistorical and unprecedented phenomenon,

it is an oxymoron,
a physicalness of expression,
just one big Walkman;
it is the most advanced form of interactivity to date,
a bioapparatus,
a form of integration of the senses,
an antidote to problems we have not solved in the world,
the solution looking for a problem;
it is a sleight of the ear, the eye, the hand, the body,
an illusion.[16]

A psychoanalytic, yet simplistic, description of a city would be a place where people live, bound together by their fears, their hopes, and their innate familial impulses. At the very least this place houses all the technical means the citizens can employ—from systems of thought to systems of manufacture—to promote their survival. Perhaps the citizens of this city want survival to seem worth promoting, and so they fabricate ideas, objects, places, and events that give value to themselves and their city.

Suppose a certain such city exists in North America (which includes Canada) and suppose the year is 1996 and a monument is to be built in the center of this city to commemorate the city's continuance into the twenty-first century. Suppose this monument is to be an artwork—a virtual environment artwork funded cooperatively by the city government and the area's most prominent corporations.[17] The monument, if not explicitly, certainly will implicitly be expected to perpetuate hope.[18]

I will return to the problem of hope later, but for now the relevance of psychoanalytic theory to this city's proposed monument is by no means obvious.

Unlike the ordinary scientist,[19] the psychoanalyst depends upon an oral and written tradition that is not primarily based on measurement but, rather, is centered upon hypotheses, observations, and discussions.[20] And, as I have already said, this city's situation requires an abnormal psychoanalytic theory, a distorted theory—distorted by the demands of culture that at times experiences dilemmas not reducible to technical quandaries.

The cultural desire in our supposed city is that this monumental virtual environment be fabricated to "mean something." The political question is "mean something to whom?" The object relations question is "does this monument imply that our lives are valuable?" The political counterquestions are "whose lives, valuable to whom, and by what criteria?"

The government and the donor corporations will not readily proceed in a way that contradicts their vested interests. If, surprisingly, their vested interests overlap exactly with those of each and every citizen interpreting the monument, then I am telling the story of a city that is, whatever else it may be, very stable indeed. In such a stable city, the monument can with conviction be fabricated to embody the relationship between the government, the corporations, and the citizens; it can remind all three parties that their past, their present, and now their future together is, was, and will be valuable to all of them.

Ambivalence, not to mention conflict, among these three parties requires different methodologies of representation[21] of the past, the present, and the future. The monument cannot be fabricated to embody a common value or meaning. Politically, one must demand different methodologies of representation.

Psychoanalytic theory can only investigate what it is to *desire* different methodologies.

The Psychoanalysis of an Ambivalent Bachelor

She came on the bus,
She came to the valley,
She couldn't say No.
She done him wrong,
She shall have music,
She should'a said No.
She waits.
She wore a yellow ribbon,
She'll be wearing pink pajamas,
She's back.

She's dressed to kill,
She's gotta have it,
She's having a baby.
She's in the army now,
She's out of control.[22]

Of all the citizens who might populate this hypothetical city, I can only speak for the ones who find object relations theory valuable for a discussion of art. Two aspects of virtual reality as art, then, would receive the most attention: first, so-called interactivity, and second, the computer's unlimited capacity for forgery. The relation between audience and work of art has already been termed interaction, and what virtual reality adds is an opportunity for literal interaction between a person's body and artwork. I am using the term forgery to emphasize, from the consumers' position, that we can no longer pretend there is such a thing as authentic documentation (artists have known this forever, but not everyone has wanted to believe them).

Ambivalent bachelor girls,[23] limited to object relations theory, would give very serious thought to the amenability[24] of the proposed monument, and would believe their interaction with this virtual environment is a politically and ethically charged process; their interpretations, as acts of elaboration, would not, as a process, change just because the medium in which an artwork is realized has changed.[25]

Virtual reality, both as promise[26] and as limited experience,[27] provokes ambivalence in this way psychoanalytically: although, consciously, one may be fully informed of the enthusiasm[28] and the repulsion[29] with which virtual reality has been greeted, insofar as it is a machine, metaphorically, symbolically, preconsciously, or subconsciously virtual reality can stand in for technology.[30] Virtual reality can be seen as the materialization of the reverence our society accords technology. And, coincidentally, the Technological Ethos inflames ambivalence—like all foci of power.

Not everyone who is reading this essay fancies themselves as an ambivalent bachelor girl, so I will magnify the details of this position. In the following passage from another essay, "virtual reality" or "virtual environment" has been substituted where originally the word "camera" had appeared.

> Is it possible for images to represent the virtual environment apparatus suspended between two visions of technology—the influencing machine and the roman machine?—neither of whose kind of truth can be denied? The representation of the virtual reality system as both of these at once[31] is a paradox that is possible only in "the third area"[32] of human endeavor. For my hypothesized subject, virtual environment is held paradoxically in her psyche because technology is. Yet how might the ambivalence be addressed within the generated images themselves? Is it a necessity or a luxury that the images represent the hypothesized subject's ambivalence toward virtual environment-as-machine? If the ambivalence is not present in the artist, what relations are set up between the computer-generated images and the viewer?[33]

Freudians and object relations analysts have respected the entwining of ambivalence and meaning. For Freudians, ambivalence is based on the conflict between body *function* and social regulation. Winnicott would base meaning on body *experience* (one's body cannot ever be objectively—or else purely subjectively—involved) in a variably responsive milieu.

The literal involvement of the audience's body or nervous system in an image would not *insure* that the virtual environment artwork has more potential for meaning[34] (for example, insure multiplicity of inferences and interpretations; call attention to the potential space between "me" and "not-me" in which acts of elaboration take place). One might even worry that immersive experiences (because they can potentially engage all the senses and increase the attention drawn to computer-generated stimuli) could jam-pack the so-called "third area" or "potential space." Then supposedly the sensory input could actually be experienced as persecution.

> Psychic reality and the actual world are relatively constant, one being biologically determined and the other being common property. The potential space between . . . individual . . . and the world depends

> on experiences that have [led to] trust. . . . By contrast, exploitation of this area leads to a pathological condition in which the individual is cluttered up with persecutory elements of which he has no means of ridding himself.[35]

Literal interaction between the viewer and computer-generated stimuli of virtual environments has been touted as either qualitatively or quantitatively distinct from that in other media. Proving which distinction is true has been the implicit agenda for both enthusiasts and skeptics debating the potential of virtual environment as an artistic medium. Psychoanalytically, the difference between virtual reality and other media would be quantitative, not qualitative.

> Viewers who cannot contribute subjectively to the interpretation as well as perception of a phenomenon cannot be in a position of responsibility, but must either submit to the authority of the work, or attempt to dominate the work by proving their power.[36]

When the audience contributes to the plot, they may change the moral of the story. Contributing one's literal nerve endings may change the intensity of the story. But having the potential and existential space in which to summon memories, beliefs, hypotheses, curiosity, and longings to link the story to what one values in one's life lived with other people is not reducible to—or exchangeable for—the moral ("the message"), no matter how intensely it is transmitted and received.

Nowadays, at the turn of the century, hopelessness takes the form of cynicism and nihilism, presumably because illusions have been substituted for objective truth, persuasion substituted for reason. Computers at last have convinced us that authentic documentation was impossible all along since documentation implies the use of a rational technique for recording objective truth. Yet when Paula Abdul cavorted with Groucho Marx and Cary Grant to promote the message that Pepsi is timeless, advertising finally managed to demonstrate what artists and psychoanalysts have wanted to bring into the public forum—that there is nothing new about the desire to forge our histories.

I want to make good my promise to return to the subject of hope. Illusion-making and acts of interpretation are still in the domain of everyone, not simply of the privileged and powerful. Psychoanalytic theory, in its limited way, can contribute to the recognition of the very high value given to illusion and to our powers of interpretation, a value that was not in any way fathered by the Technological Ethos.

Notes

1. Jeanne Randolph, "Influencing Machines: Undone," *Span, L'Esprit Nerveux* (Spring 1992), 34.

2. "To conceive of technology as a representation is to bring it into the realm of metaphor (albeit the dominant metaphor of the human predicament as depicted by capitalism and the military). . . . The vulnerability of a metaphor that is seldom acknowledged in regard to technology is, however, that metaphors evoke not only an objective but also a subjective response, and the latter response signals that perhaps one should be vigilant, a keen observer, judgmental. A metaphor invokes evaluation of its claims." Ibid.

3. "Technology could . . . be thought of as an ideology in which human experience is defined and perceived exclusively in terms of goals to be achieved as efficiently as possible. The ideal . . . would be to find 'the one best way' to accomplish each goal. Thus technology will not concern itself with the experience of the subject, but instead with the materiality of the object." Jeanne Randolph, "Influencing Machines: The Relationship between Art and Technology," in *Psychoanalysis & Synchronized Swimming and Other Writings on Art* (Toronto: YYZ Books, 1991), 38.

> One comes to learn a procedure by which, through a deliberate direction of one's sensory activities, and through suitable muscular action, one can differentiate between what is internal—what belongs to the ego—and what is external—what emanates from the outer world. In this

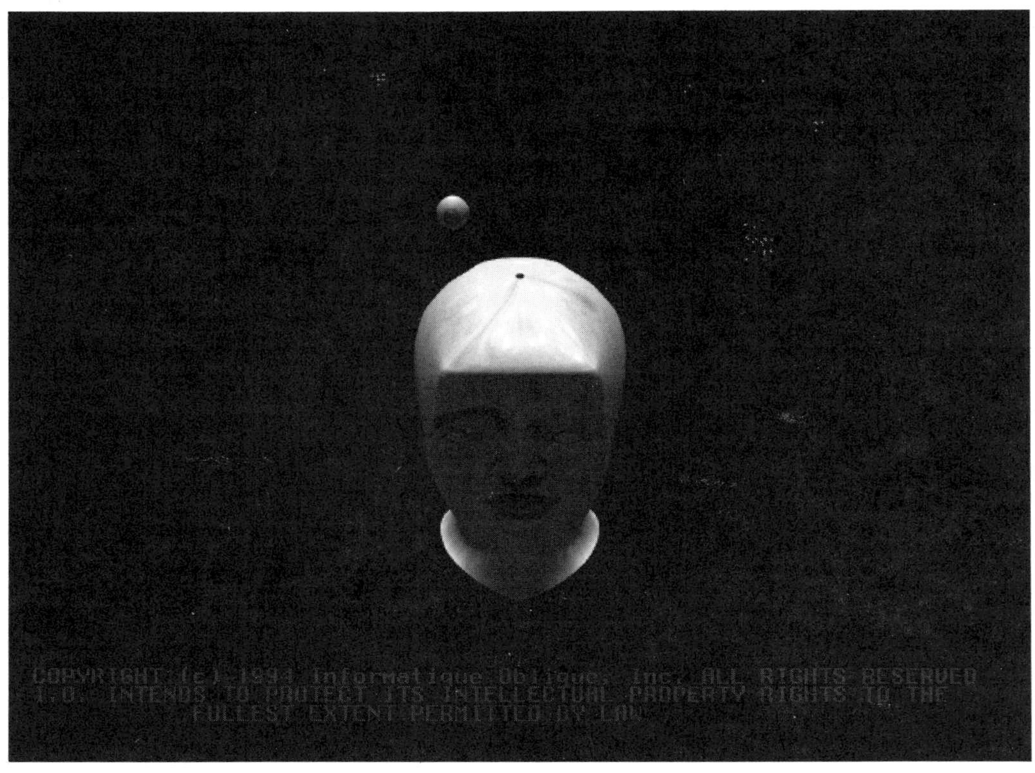

18. Ron Kuivila, *VR on $5 a Day* (1994), frame grab of portfolio personification.

> way one makes the first step towards the introduction of the reality principle, which is to dominate future development. This differentiation, of course, serves the practical purpose of enabling one to defend oneself.

"This proclamation from Freud's *Civilization and Its Discontents* depicts the human subject in relation to the reality principle. To date, in my thinking, I have not got around to making a distinction between this Reality Principle and the Technological Ethos." Jeanne Randolph, "Culture and the Technological Ethos," paper presented at the MacDonald Stewart Art Gallery for the University of Guelph "Ethics and Technology," international conference, October 1989.

"Technology's only reality is implementation in the active, literal sense. To insist that what is subjectively true can be absolutely separable from what is objectively true is intrinsic to the Technological Ethos—and constitutes its brutality." From "Technology as Metaphor," in *Psychoanalysis & Synchronized Swimming,* 77.

"The Technological Ethos tends to diminish paradox, diminish subjectively distorted experience, and give no recognition to the impulse to interact with society for other than physical, social or economic self defense. . . . It is pervasive, because everything we want requires some technique to obtain." From "Culture and the Technological Ethos."

"The prevailing fantasy that arises out of the Technological Ethos, in which efficiency and objectivity are the supreme values, is the fantasy that any facet of life can be transformed into a world class event, if it happens clean enough, fast enough and aggressively enough." Jeanne Randolph, "Robin Collyer Exhibition at Carmen Lamanna Gallery," *Parachute* 42 (1986), 41.

"It is not the existence of the Technological Ethos that is cause for alarm. It is its absolute *dominance.*" From "Culture and the Technological Ethos."

4. In an interview in *MV5* (Winter 1991–1992), 25.

5. Judith S. Donath, "Identity and Deception in the Virtual Community," paper presented at the Fourth International Conference on Cyberspace, Banff Centre for the Arts, May 1994, 26.

6. Brenda Laurel, *Computers as Theater* (Reading, Mass.: Addison-Wesley, 1991).

7. "We may expect that one day someone will venture to embark upon a pathology of cultural communities," said Sigmund Freud in 1939 in *Civilization and Its Discontents,* noting that "the present cultural state of America would give us a good opportunity for studying the damage to civilization to be feared." *Civilization and Its Discontents,* ed. and trans. James Strachey (New York: W. W. Norton, 1962).

8. Depicting a society as diseased has been irresistible for a very long time. For instance, Ezekiel 34:1–4 transmits God's reproof to the Israelites for living "with force and with cruelty" such that their shepherds are obviously at fault for "the diseased" among them. The shepherds have not "healed that which was sick, neither have ye bound up that which was broken . . ." (quoted from the King James Version of the Bible [Chicago: John A. Hertel Co., 1938], 890). More recently, Arthur Kroker, for example, claims "America is perfectly schizophrenic" (in *Mondo 2000* [1994], 65).

9. "What relation does a feminist establish with a project such as psychoanalysis, a body of thought that is historically legitimated, patriarchal and surrenders itself to bourgeois capitalist sentimentality? By any interaction with theory whatsoever, am I not balancing between folly and bravery, analogous perhaps to sculpting marble in the era of virtual reality? Yet there must be a way to deal with this historically legitimated, patriarchal bourgeois capitalist theory and squeeze out of it something that is relevant and timely. I will emphasize the word squeeze, if you will allow me this indulgence: psychoanalytic theory is the medium in which I work. I find some way to engage with it, distort it, so that it is no longer exclusively a patriarchal, scientific, objective methodology. This would be my Playdoh hypothesis of how psychoanalytic thought works. I am trying not to proclaim psychoanalytic theory, but rather to monkey with it. And that is the utopian impulse available within analytic theory, to wish that distorting theory will show its limits and also its relevance." Jeanne Randolph, "Technology and the Preconscious," in *The City Within* (Banff: Banff Centre for the Arts, 1992), 36–37.

10. Jeanne Randolph, "The Amenable Object," in *Psychoanalysis & Synchronized Swimming,* 31.

11. "A two-part definition of primary process will be offered: there is first of all what primary process does and second why it does it. What primary process does is to operate by way of disregarding logic, disregarding reason, disregarding practicality, ignoring the

physical and political restraints of day-to-day action, interchanging basic intrinsic qualities of things with trivial or incidental qualities of things, rearranging and regrouping qualities of things, parts of things and whole images of things to condense them into symbols. Secondly, why it does this, is that the sole aim of primary process is the discharge of any accumulated erotic or aggressive arousal. Through displacement, symbols become the means by which this excitation is relieved. When intrapsychic quiescence has been re-established that is, by definition, pleasure, the pleasure that is intrinsic to the subconscious." Randolph, "The Amenable Object," 24.

12. According to theorist John Hunter Padel, Melanie Klein (1882–1960) "put the relationship of feeding mother and infant at the centre of the picture; later relationships came to be interpreted in terms of that one. . . . D. W. Winnicott (1896–1971) saw attachment-bonds being developed and then becoming loosened (he never lost sight of the idea that for the infant at first his mother has a dual role; she is his environment and she is his object [of desire]). . . . The child at every stage chooses between living in his own illusions and acknowledging [the mother's] provision for his needs." John Hunter Padel, "Freudianism: Later Developments," in *The Oxford Companion to the Mind* (London: Oxford University Press, 1989), 273. I have commented: "The psychoanalytic work of D. W. Winnicott, particularly as expressed in his book *Playing and Reality,* raises the possibility that in art it is the ambiguity between the objective and the subjective that gives artwork a unique psychological validity. Through Winnicott's observations and speculations, the model of the art object is of an object amenable to an interaction with the viewer, reflecting the hypothesis that in some way the material and methods of which it was made had been rendered by the artist into something amenable to subjective interventions, a subjectivity very like primary process, yet exploratory, not always reactionary." (For a detailed discussion of the relevance of Winnicott's thought to the process of writing art criticism, see my "The Amenable Object," 21–35.)

13. Although I have a modicum of autobiographical insight into my choice of object relations theory as the medium in which I work to interpret the North American predicament in which I live, I cannot pretend to be objective about my own commitment. Object relations theory is valuable to my work because it recognizes binary thinking as a rhetorical device that *creates* paradoxes and self-contradiction; this theory recognizes the absolute necessity of mothers (regardless of their gender); object relations theory is so ambiguous that it is useless without deliberate reflection, tinkering, and speculating.

14. In Winnicott's view, "in regard to the importance of good-enough mothering . . . this includes fathers, but fathers must allow me to use the term maternal to describe the total attitude to babies and their care." D. W. Winnicott, *Playing and Reality* (New York: Pelican, 1985), 166. Winnicott speaks also of the "mother-figure" when the biological mother is not the primary caretaker (ibid., 55).

15. There are two processes that object relations theory defines as intrinsically cultural: to sustain an "illusion" and to make an interpretation. Both of these concepts are extrapolations from D. W. Winnicott's clinical discussion of "the transitional object" (see *Playing and Reality,* chap. 1). For instance, rather than Freud's dualism, in which a strict distinction is made between reality and fantasy, Winnicott claims that "if there is a need for this double statement there is also a need for a triple one: the third part of the life of a human being, a part we cannot ignore, is an intermediate area of *experiencing,* to which inner reality and external life both contribute. . . . I am staking a claim for an intermediate state between . . . inability and growing ability to recognize and accept reality. I am therefore studying the substance of *illusion*. . . . We can share a respect for *illusory experience*" (ibid., 3). Illusion delineates a zone between pure mental states: between a state of delusion on the one pole and of objectivity on the other ("there are [people] who are so firmly anchored in objectively perceived reality that they are ill in the opposite direction, of being out of touch with the subjective world and with the creative approach to fact") (ibid., 78). In reading object relations theory for its implications for culture, I have taken Winnicott's phrase "what is objectively perceived is by definition to some extent subjectively conceived of" and speculated upon the colonization by the Technological Ethos of those aspects of the *audience's* psyche that "subjectively conceive of" (co-create, embellish) artworks. This has resulted in my discussion of interpretation as "an act of elaboration" that may or may not be dominated by the Technological Ethos. Another way to depict what I have called "an act of elaboration" is to revise Coleridge's phrase "the willing suspension of disbelief" to include "the conscious or subconscious contribution of additional belief."

16. This is a list of all the depictions that begin with "virtual reality is . . ." in order of appearance in the proceedings from the Virtual Seminar on the Bioapparatus. See *Virtual Seminar on the Bioapparatus,* ed. Catherine Richards and Nell Tenhaaf (Banff: Banff Centre for the Arts, 1991), 8, 10, 10, 16, 17, 26, 36, 45, 48, 62, 63, 64, 79, 87, 90, 95, 97, 99, 100, 109, 111, 114, 114.

17. The constraints of this situation are consistent with "the location of cultural experience" in object relations terms: "I suggest that the time has come for psychoanalytic theory to pay tribute to a third area between objective reality and subjective fantasy, that of cultural experience which is a derivative of play. . . . I have located this important area of experience in the potential *space between* [italics mine] the individual and the environment. . . . Maximally intense experiences [of a citizen, for example] take place between the subjective object [the virtual environment, for example] and the object objectively perceived [that same virtual environment]. . . . Maximally intense experiences take place . . . at the interplay between there being nothing but me and there being objects and phenomena *outside my omnipotent control* [italics mine]. . . . In using the word culture . . . I am thinking of something that is in the common pool of humanity, into which individuals and groups of people may contribute, and from which we may all draw *if we have somewhere* [intrapsychically, interpersonally, and into our praxes] *to put what we find*" [italics in original]: see Winnicott, *Playing and Reality,* chaps. 7, 8, pp. 112–129. "Reshaping Winnicott's theory to the aims of art criticism allows a way to interact with the artwork as an intentional revelation of the artist's version of experience, intentionality that need be neither explicit nor disguised. . . . This is not a theoretical model of the artwork from the purportedly objective view of someone who wants to study how the artist creates it. This is a view of the art object once the artist has left it in public." (Randolph, "The Amenable Object," 31.)

18. And this is not to say that the artist(s) who produce(s) the monument will comply with this expectation.

19. When ordinary scientists propose an investigation or experiment according to the scientific method, they will include: (a) the theory and history of previous inquiries of the phenomenon under consideration; (b) the hypothesis to be tested; (c) the materials and methods to be used; (d) the observations made and/or the objective data collected; (e) the interpretation of these results; (f) the conclusions that can be drawn, including the status of the original hypothesis; (g) a discussion of alternate interpretations of the data or other conclusions that seem however to be less likely; and (h) implications for further investigations.

20. Freud has his own view: "Psychoanalysis, in my opinion, is incapable of creating a *Weltanschauung* of its own. It does not need one: it is part of science and can adhere to

the scientific *Weltanschauung*. . . . A *Weltanschauung* erected upon science has, apart from its emphasis on the real external world, mainly negative traits, such as submission to truth and rejection of illusions. Any of our fellowmen who is dissatisfied with this state of things, who calls for more than this for momentary consolation, may look for it where he can find it." Sigmund Freud, "The Question of a *Weltanschauung*," lecture 35 in *New Introductory Lectures on Psychoanalysis* (New York: Penguin, 1944), 219.

21. It is my choice to pretend that in the city where the virtual environment will be built, there is ambivalence and conflict. The artist(s) who produce(s) this monument may or may not be consciously ambivalent about the conditions given.

22. A list, in alphabetical order, of movie titles beginning with "She . . ." from *Videohound's Golden Movie Retriever*, ed. Martin Conners and Julia Furtaw (Detroit: Invisible Ink Press, 1994), 740–743.

23. Despite all the connotations of the word *bachelor*, it has only a few synonyms—single man, unmarried man, "bach," or "old bach." There are more synonyms for an unmarried woman: single woman, unmarried woman, spinster, old maid, maiden, virgin, bachelorette, and bachelor girl. Akin to offering the feminine version of the oedipal complex, I presume to offer one possible feminist version of a bachelor machine. Unlike Freudians, I am under no obligation rhetorically to invent a bachelor girl machine that is the exact opposite of, or corresponds element-by-element to, the supposed masculine version.

24. "The amenable object was my attempt to avoid conceptualizing interpretation as an hermeneutic act determined by Freudian dualism. . . . In his model, illusion becomes the *opposite* of the real. . . . The kind of illusion I am referring to when I speak of a symbolization, a representation, takes place through another kind of interaction besides expert deciphering. A found object becomes what I term an illusion only if the audience agrees to sustain it and then provides something for it: an interpretation ('an act of elaboration'). . . . It seems some of my acts of interpretation of found objects are quite deliberate, and I take conscious pleasure in my multifarious elaborations. And some of my interpretations take place in a millisecond, preconsciously, and I am hardly aware of the part I played in establishing the illusion." (Randolph, "Culture and the Technological Ethos.") "As an amenable, rather than a utilitarian object, the artwork itself never

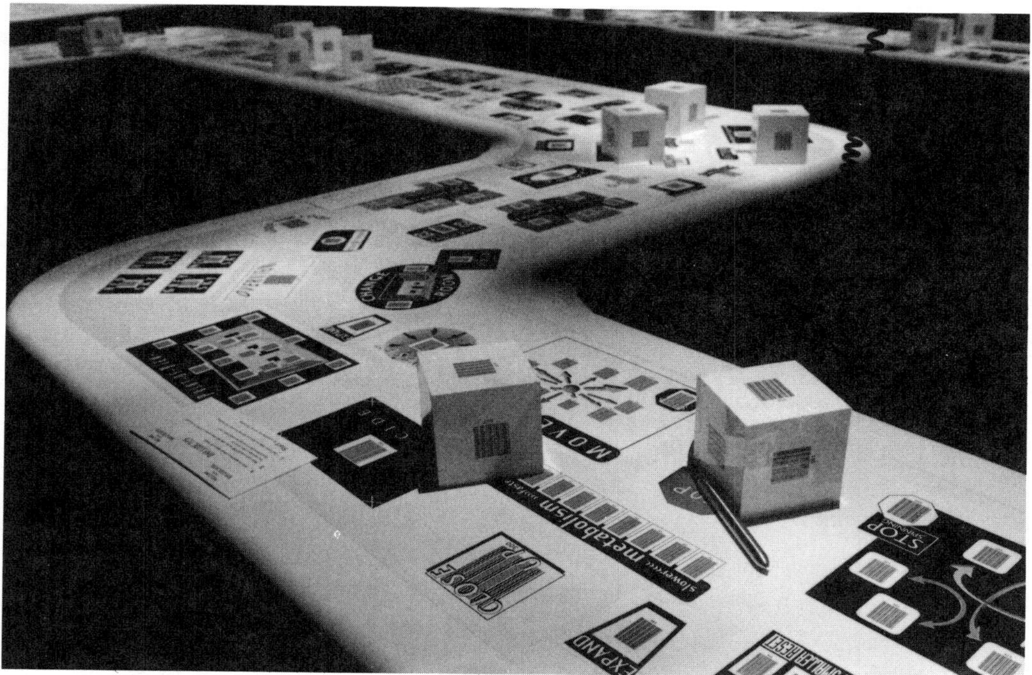

19. Perry Hoberman, *Bar Code Hotel* (1994), installation detail. Scanning the bar codes on the cubes with the wand will bring objects into existence. Scanning bar codes on the walls and table tops will modify their behaviors. Photo Donald Lee.

acquiesces to any mutually exclusive polarities it seems to manifest. A collegial interaction, in which the viewer contributes equally to meaning and interpretation, is made possible, rather than a relationship in which artwork is an authority and the audience a voyeur." (Randolph, "Influencing Machines: The Relationship between Art and Technology," 53.) "I am imagining a subject who looks at an artwork not because she is consciously or unconsciously seeking resolution or quiescence, nor because she is seeking in vain a proof of her own coherence. This is a subject who welcomes the conditions in 'the third area,' conditions that are ambiguous, self-contradictory, approximate. These conditions necessitate an act of interpretation. By an act of interpretation this hypothesized subject deliberately co-authors illusions so typical of a realm between the reality principle and the pleasure principle, illusions in the sense that the hypothesized subject ponders, but leaves unspecified the extent to which the illusions are both present objectively and presented as metaphor." (Randolph, "Technology as Metaphor.")

25. "Implements, devices, functional things, from guns to butter, would by the object relations definition of illusion, remain literal objects until someone presented them as a metaphor to someone else who was in a position to toy with the metaphor. . . . At the moment when I know I am the one wielding the power to interpret an object, when I find the interpretation more valuable to me than the function that the object serves, at that moment the object could become cultural." Jeanne Randolph, "Illusion and the Diverted Subject," in *Psychoanalysis & Synchronized Swimming,* 68–69.

26. Theoretically, it is about time that art's contribution to "a radical transgression of the boundaries between masculine and feminine, human and alien, human and machine, life and death, time and space" (in the words of Inez van der Spek, in "Transcendence out of or into the Body," in *Virtual Seminar on the Bioapparatus,* 59) is addressed as an inevitability in all cultural forms, but it seems to have taken the invention of virtual reality to accentuate this aspect of symbolization. "From my perspective," writes Doug Hall ("Concerning Virtual Reality and the Bioapparatus," in ibid., 87), "it is more interesting (and I think more productive) to be a critic and observer of media, to use tools [especially VR] as part of a larger aesthetic, conceptual and critical endeavour, than it is to fetishize the tools for their own sake." Like Itsuo Sakane, I am arguing for "a balance between two types of art in this age: a harmony is possible in our personal and social lives between the art of synthetic reality, based on information technology, and

the art of the five human senses, based on material." (See his editorial in *Leonardo* 24 [1991], 3.) From this standpoint, "phenomena that are brought into the intermediate or third realm of activity no longer exist solely on their own terms. In the cultural realm the objective and the subjective merge. A judgment regarding the value of phenomena subjected to this blurring within the cultural area is returned to the people affected by the phenomena, who can now interact with the latitude to revise the phenomena perceptually rather than merely react with complicity or dismissal. In this third, cultural realm of activity, phenomena become found objects that are free to be redefined as well as actually reshaped, so that their interpretation is multiplied." (Randolph, "Influencing Machines: The Relationship between Art and Technology," 48.)

27. This essay results from firsthand participation in or viewing of: Michael Scroggins and Stewart Dickson's *Topological Slide,* Lawrence Paul Yuxweluptun's *Inherent Rights, Vision Rights,* Michael Naimark's *See Banff!,* Perry Hoberman's *Bar Code Hotel,* Toni Dove and Michael Mackenzie's *Archeology of a Mother Tongue,* two versions of *Dancing with the Virtual Dervish,* one by Diane Gromala and Yacov Sharir and the other by Marcos Novak, Ron Kuivila's *VR on $5 a Day,* Will Bauer and Steve Gibson's *Objects of Ritual,* two screenings of Catherine Richards's *Spectral Bodies,* and a demonstration of some of the virtual environment components of Brenda Laurel and Rachel Strickland's *Placeholder.*

28. Warren Robinette: "The electronic augmentation of human mental and physical powers leaves our humanity, our ethics and our judgement intact and in control. One vision of the future is the human being whose senses and muscles are greatly amplified, a human decision-maker aware and powerful." ("Technological Augmentation of Memory, Perception and Imagination," in *Virtual Seminar on the Bioapparatus,* 17.) Mary Ann Amacher: "Small children, instead of having to watch television and having this mode or path of entertainment, can actually have friends all over the world." (In the discussion after "Designing the Social," in ibid., 34.) Fred Truck: "Virtual reality technologies will allow the user to experience physically, not only the actualized vision of a great artist of times long gone, but the intense and vivid dreams many have of flying, swimming through the air, or blasting through the coronasphere to a new day, without falling." ("Art Machines," in ibid., 69.)

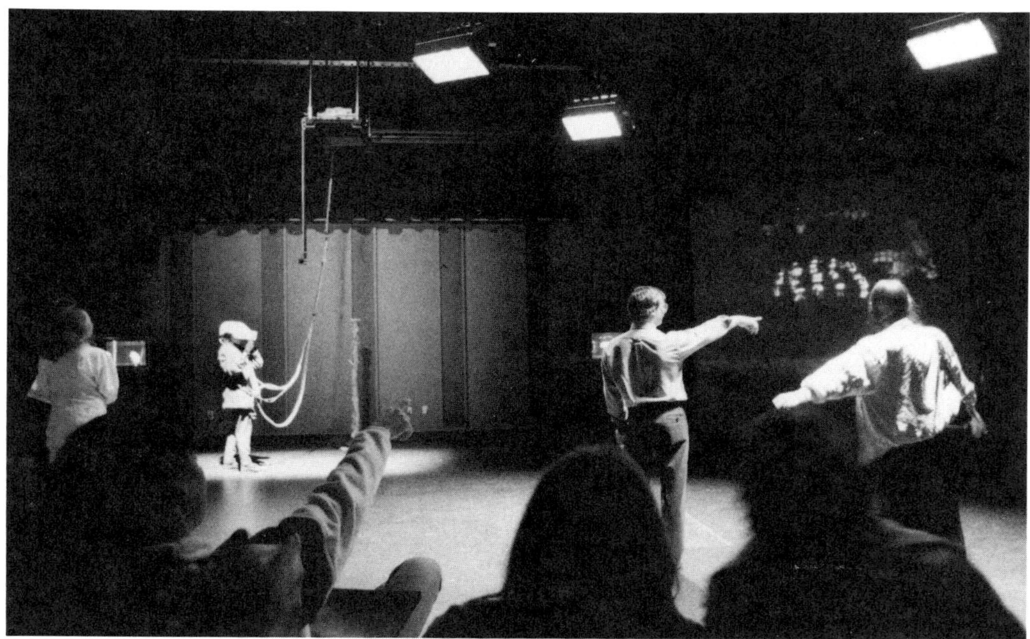

20. Ron Kuivila, *VR on $5 a Day* (1994), installation view showing participants interacting with the virtual environment in the installation space while the "helmethead" in the background is immersed in the environment. Photo Donald Lee.

29. David Rokeby: "Mind and body annihilate each other, leaving only the evolutionary potential of technology." ("Evolution and the Bioapparatus," in ibid., 18.) Lyne Lapointe and Martha Fleming: "Virtual reality heightens and maintains a polarity wherein there is no pleasure to be got from understanding, and there can be no understanding of our pleasure." ("An Iron Hand in a Velvet Glove," in ibid., 38.) In response to the Bioapparatus seminar, David Tomas: "And I fear, as the events of the recent past are still fresh in my memory, that [VR's] cultural logic, its master text, will very likely remain the same. . . . In other words, our consciousness, perhaps the last site of socio-cultural resistance, will soon become open to the index of the law, and its inscription surface will no longer be an epidemic shell—the skin—but the neural networks of our fragmented but nevertheless still gendered minds." ("Response to Subjectivities," in ibid., 62.)

30. I have previously analyzed the machine as the physical embodiment of the Technological Ethos, exploring the relevance of the psychotic delusion of "The Influencing Machine," and, at the other extreme, the relevance of the "Roman Machine," a delightful, redemptive gadget that is the figment of what Melanie Klein has termed "manic denial." (See my "Influencing Machines: The Relationship between Art and Technology," 37.) Subjectively, ambivalence about technology in its most intense form vacillates between these two extremes, each of which has its own kind of poetic truth. The Influencing Machine is a device that serves to persecute, and is perpetrated by enemies. It produces, as well as removes, thoughts and feelings; its effect is to control the content of one's inner life, to control the subjective privacy ordinarily experienced as impenetrable except by invitation. There is a progressive distortion undergone by this machine and consequently, eventually, it loses all characteristics of the human body and becomes unintelligible. The original essay about this delusion was written in 1933 by psychoanalyst Victor Tausk ("Origins of the Influencing Machine in Schizophrenia," *Psychoanalytic Quarterly* 2 [1933]). "The Roman Machine . . . would remain entirely innocent, guiltless. It serves to delight and is operated by friends. The effect of the Roman Machine is to reassure that physical, intellectual and emotional pleasure belong entirely to oneself. The distortion that this apparatus has undergone is that it has lost all relevance to the choices humans must make in socio-political life, i.e., lost all relevance to the body politic." The original essay on which this is based was written by psychoanalyst Hanns Sachs ("The Delay of the Machine Age," *Psychoanalytic Quarterly* 2 [1933]).

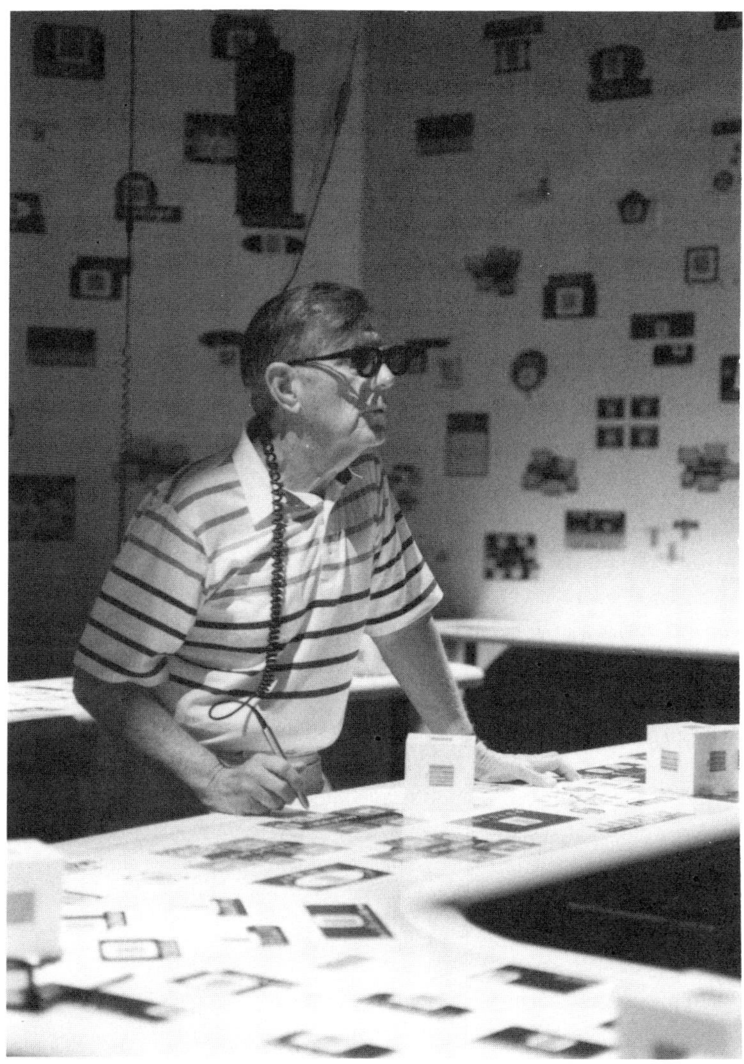

21. Perry Hoberman, *Bar Code Hotel* (1994), installation detail. Each participant has his or her own wand for scanning objects and manipulating them. Photo Donald Lee.

Perry Hoberman's *Bar Code Hotel* immersed participant(s) in what might have seemed at first impression to have been a Roman Machine. It was possible to enjoy this virtual environment unselfconsciously, as did all of the children that I saw when it was installed at the Banff Centre in 1994, who adored it. The images, the text, as well as one's own muscular and virtual actions were extremely stylized—in the tradition of the cartoon from Fritz the Cat to Ren and Stimpy—so that the status of images and texts as pure representation was totally obvious. None of the "characters" (a hat, a pair of shoes, a radio, sunglasses, for example) were depictions of a life form, however. There were no bunnies, gnomes, no robots, not even insensate organisms such as protozoa or beans. All of the "action" accessed through the bar codes was charmingly oversimplified by text commands, such as "jitter," "punch," "breathe," as well as more pragmatic verbs such as "expand," "contract," "avoid," and "merge." One's absolute command of one's icon (or of brief "events" such as earthquakes, for example) was by no means assured. Some autonomy had been programmed into each icon. "Perhaps the best analogy," according to Hoberman's sense of these responsive icons, "would be that of an exuberant, misbehaving pet." When as many as a dozen people would be involved, the entire field of imagery and the effervescent, rudimentary narratives generated were unpredictable. Delusions of ecstasy, freedom, and omniscience with which virtual reality has been promoted were conspicuous by their absence.

Bar Code Hotel also plays with the ever worrisome distinctions made between so-called high and low culture, between art and entertainment. Even if *Bar Code Hotel* was built to more closely resemble so-called entertainment (and this could be argued psychoanalytically only on the basis that it did not evoke memories, free association, ambiguity, or guilt) than so-called art, nevertheless, interacting within the space of the *Bar Code Hotel,* I did not become "so enfeebled by the conditions set up by this divertisement that I am ever more willing to abdicate my role as co-author of illusion" (Randolph, "Illusion and the Diverted Subject," 68).

31. "A parody of the evolution of virtual environments—it's a suicidal idea," said Ron Kuivila, talking about how he might address the contradictions in the development of virtual technologies. His work *VR on $5 a Day* was experienced as a series of encounters with computer-assisted imagery and soundscape, some connoting and some denoting the various parties who have a vested interest in this technology. The audience participated as spectator(s), "helmet-head(s)" (Kuivila's term), or activator(s) of icons on a screen.

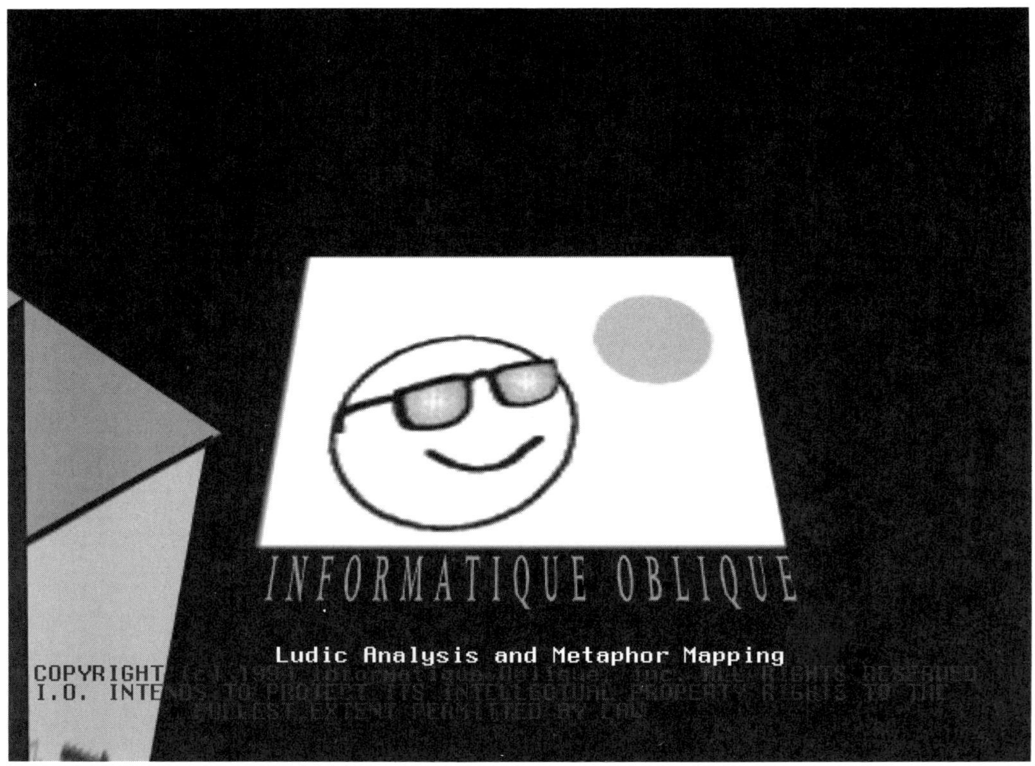

22. Ron Kuivila, *VR on $5 a Day* (1994), frame grab from one of six advertisements. This one shows the rose-colored glasses of metaphor mapping.

However one chose to participate, there was no control over loud interruptions by garish commercials (one of which was a kind of outrageously useless public service-type fitness break). The following is an excerpt from the Metaphor Mapping® ad, designed visually to be so clean and slick (as were all the "ads") that it could only have been the product of anal retentive psychosis—or of an artist, in this case, who appreciated the excesses of advertising:

> The result? Metaphor Mapping® technology, the root of our corporate tree. Simply put, Metaphor Mapping allows any problem or task to be re-expressed as an interactive game. Once ludified, the problem is posted to our intelligence Mine, where millions of crack game players, our Intelligence Miners, can have a go at it. As the I-miners play, winning strategies appear and our automated Ludic Analyzer, Laszlo, collects the strategies for deludification and application.

Serene, rather pretty interludes were also presented if one donned the helmet and glove. These were fragile, brief, and unproductive lapses, like star-gazing in one's backyard, so benevolent one could almost believe virtual reality is a roman machine. But soon enough the participant is delivered again to the actuality of corporate dominance.

32. "D. W. Winnicott . . . [emphasized] the interaction between the artist and the external world. . . . Unlike Freud and Klein [he] did not discuss this interaction, or the artwork that embodies it, using dualities such as inside versus outside, fantasy versus fact, intuition versus reason. Instead, he asserted that there is an intermediate kind of human activity or experience in which there is a deliberate blur of the distinction between these. That third area [is] between autistic, private fantasy and public, pragmatic convention . . . between bodily function and sociopolitical function." (Randolph, "Influencing Machines: The Relationship between Art and Technology.")

33. I've been wondering about this for some time; see my "Technology as Metaphor," 80.

34. Meaning, especially before it is organized consciously as words, is generated in the form of memories and free association. For Freudians, meaning, insofar as it is the basis of interpretation, depends on memories, whether biologically, interpersonally, or culturally derived. Kleinians claim that something is meaningful to the degree that it

activates bonding impulses, sadistic impulses, guilt, grief, and reparation. If free association does not link images or sensations according to these essentially social emotions, the images are irrelevant. See Geraldine Pedersen-Krag, "A Psychoanalytic Approach to Mass Production," *Psychoanalytic Quarterly* 20 (1951), 414. Pedersen-Krag discussed the need for *meaningless* activity and its implications in a technological society. Meaning, then, is supplied to a certain extent by the subject; meaning is implied, not supplied entirely by the artwork.

This latter conjecture about the interaction between audience and art object accounts, at least psychoanalytically, for the intense reactions anecdotally reported in participants in *Objects of Ritual* by Will Bauer and Steve Gibson. In this context I can only report my own reaction. In retrospect I renamed this virtual environment "Virtual Reality in America." It was a multimedia extravaganza that illustrated many critical insights about the relation between the computer's capabilities and art (and, I have claimed, therefore, the relation between the Technological Ethos and culture). Psychoanalytically one could claim that this virtual experience was the subjugation of the audience to a thesis on epistemology, omnipotence, the social impulse and scopophilia. *Objects of Ritual* was an experience in which any bodily gesture or vocalization I might have initiated spontaneously would not have supplied anything of meaning to my—or other participants'—interpretation of this work. This is not to say that *Objects of Ritual* was meaningless, but rather that it was oversupplied with meaning in the hermeneutic sense of the word. Intellectually, one can infer no greater condemnation of a medium than this—the medium places its audience in a position of total sensory absorption and yet total subjective irrelevance.

35. Winnicott's words, in *Playing and Reality,* 121.

36. The phenomenon must be amenable to acts of elaboration. See Randolph, "Influencing Machines: The Relationship between Art and Technology," 51.

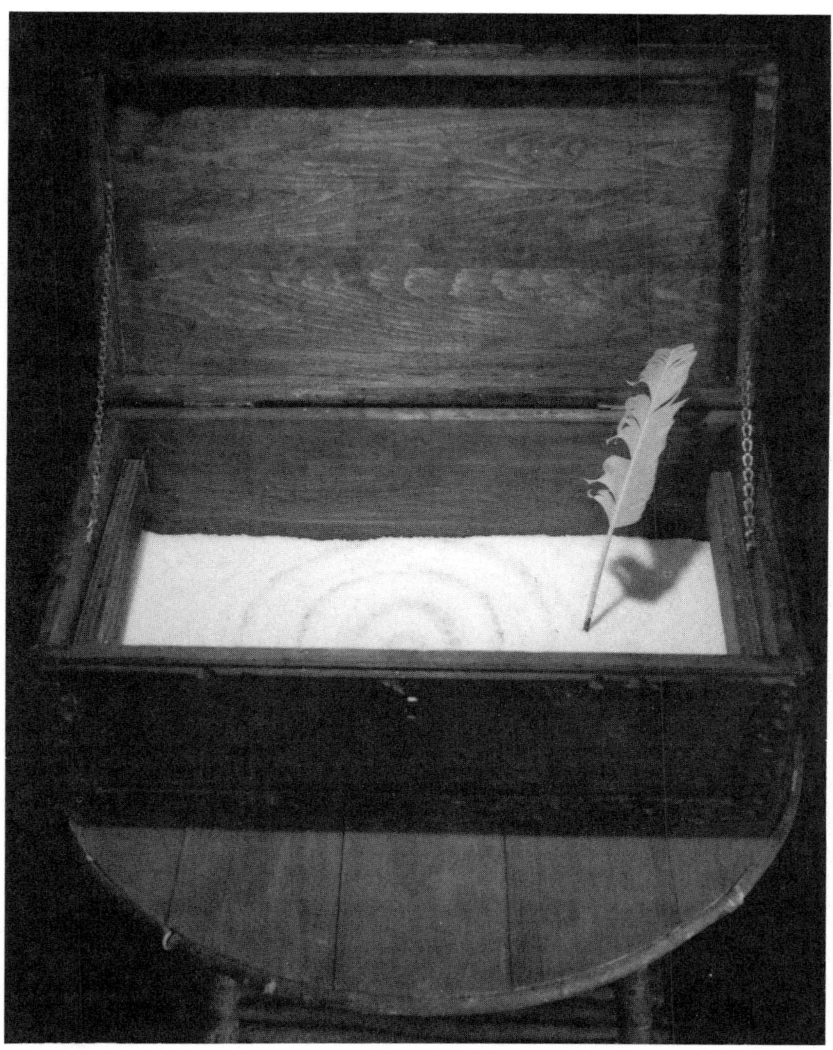

23. Will Bauer and Steve Gibson, *Objects of Ritual* (1994), installation detail. Photo Cheryl Bellows.

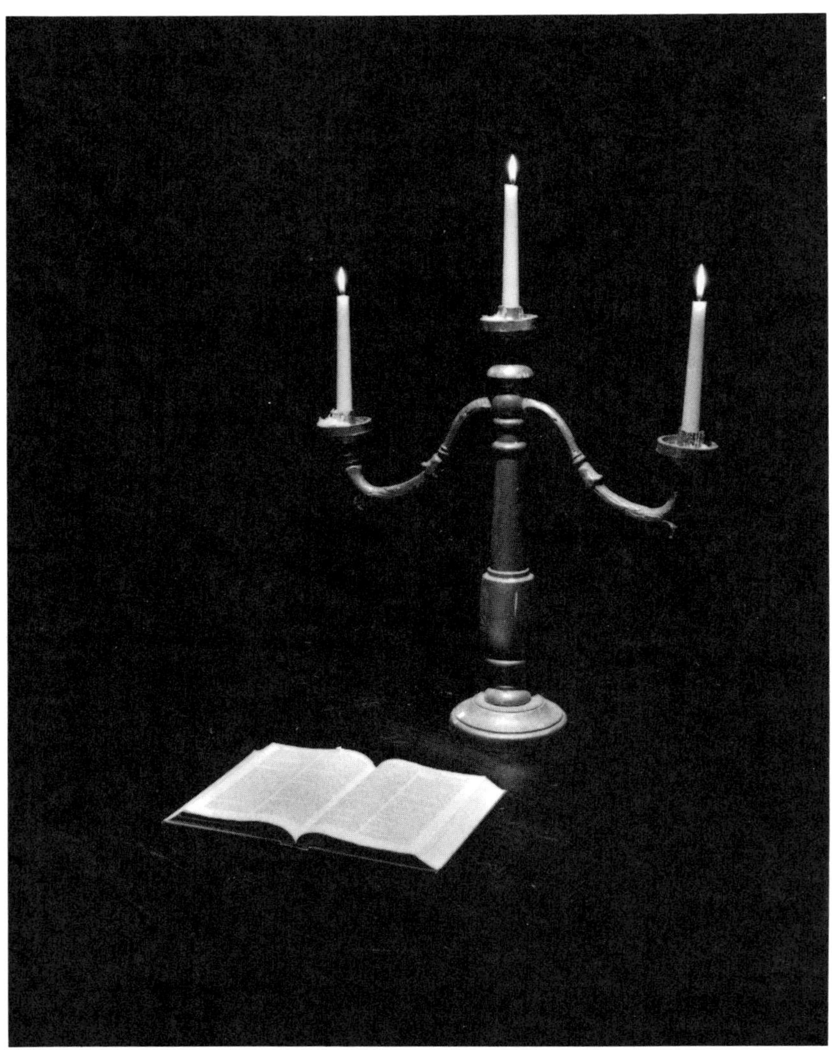

24. Will Bauer and Steve Gibson, *Objects of Ritual* (1994), installation detail. Photo Cheryl Bellows.

Aboriginal Narratives in Cyberspace

Loretta Todd

When Crees speak of Canada, they mean Ka-Kanata-Aski, "the land that is clean." To be Cree is to be Nee-yow, "We who are members of that Nation of people who are part of the four seasons."[1]

The word Ka-Kanata signifies Cree narrative history and philosophy about "the clean land" and means more than an absence of garbage.

A clean land. Clean, not as synonymous with pure, as in the Garden of Eden, but clean meaning a balance and harmony amongst "All the Relations" who share the land. A clean land that the people must protect.

What then does the new territory called cyberspace mean to aboriginal people? If it is really a place, albeit under construction (and which at the same time will not really be there when it is there), what can we name it? Can we name cyberspace the way we name other technologies, from sewing machines to chain saws? Can it, like television, be called "a talking box of space"? Will we name cyberspace a talking web of clouds? A land of mists where shadows chase shadows or a dream world with no memories or traces of before?

Can our narratives, histories, languages, and knowledge find meaning in cyberspace? And above all, can cyberspace help keep Ka-Kanata a clean land?

For instance, will the need for fossil fuel for endless commuting be reduced as Corporate Virtual Workspaces are created in cyberspace? Or will CVW alienate humanity even more from the land, allowing as yet unheard of ecological abuse?

Will cyberspace enable people to communicate in ways that rupture the power relations of the colonizer and the colonized? Or is cyberspace a clever guise for neocolonialism, where tyranny will find further domain?

Will cyberspace enable old knowledge to be experienced and expanded, or will cyberspace create the present anew each day, so that there never again is tradition or a past?

In other words, what ideology will have agency in cyberspace?

Some might say that cyberspace marks the end of ideology, the end of history. In cyberspace everyone will have free and equal access regardless of origins. Difference will be mutable, no one group will prevail because, after all, you can be anything or anyone in cyberspace.

But will cyberspace duplicate what already exists—with virtual malls, as constructed by the world of commerce, and virtual museums, as constructed by the academy, or virtual arcades, as constructed by the entertainment industry? Remember, cyberspace started as a virtual war zone, as constructed by the western military.

Much has been written about the origins of cyberspace, from the psyche that imagined it to the technology that has created it. Those origins are important to examine when talking of creating aboriginal meaning in cyberspace.

Some have said cyberspace is "Platonism as a working product." Here, "with an electronic infrastructure, the dream of perfect *forms* becomes the dream of in*form*ation. . . . Filtered through the computer matrix, all reality becomes patterns of information."[2] Here, the idea of the universe as imagined by Plato takes shape.

Perhaps these patterns in cyberspace will be imagined as liquid light, like a neon sign shining slick and menacing by a wet road at midnight. Maybe cyberspace cities will be built as if of crystals that shimmer in a sunless sky. And along the roads and highways of this geography, information will flow like blood, in a space where there will be no flesh and shapes will shift like shadows on the wall.

Propelling humanity to this place, to the perfect *forms,* is Eros "to extend our finite being . . . [and] to see more and to know more deeply." And to do so, Eros "inspires humans to outrun the drag of the 'meat'—the flesh."[3]

In this ontology of cyberspace—hell of western thought—the tension between the need to know all, to emulate *visio Dei,* and the limitations of the body and the senses, of the physical world, creates a need—for a new site for the "heart and mind" of man (although in the *Terminator* films the "heart and mind" of man was reborn in the machine, yet the machine hated humankind and sought its destruction).

Western culture is not the only one that seeks to know. The quest to find the essence, the unifying structure, "the mind of God," is shared across cultures. In native culture we too have stories in which our Prometheus—a trickster—seeks out fire or the sun, tricking his father, or a great chief, or an old woman, to give up the flame or light so that the people can have warmth and see the day. In other words, our cultures are dynamic and also seek knowledge, seek understanding (and, in contrast with Judeo-Christian belief, knowledge is not seen as potentially dangerous). Wesucechak, the Cree trickster, is, after all, always going walking, always on a journey toward understanding. But there is no Platonic effort to separate ideas from the world or, like Descartes, to separate the body from the mind. Nor is there a need to be like God.

Instead, all in the universe is endowed with spirit and intelligence, from which ideas flow. It has become almost trendy to mention, but in this world view there is no separation of body, mind, spirit, and heart. The body is of the spirit—the mind of the heart. I may stretch the credibility of the more scientific-minded, but imagine our view as resembling string theory: All life is connected throughout the universe. Yet we are of the earth, responsible for our existence in our territories, responsible for caring about the life around us. Though life is hard, there is joy. The land is the culture, the old people say—in contrast, even opposition, to that which says the mind is the culture.

It is not so odd, then, at this stage of late capitalism in the project called western culture, that cyberspace is "under construction." It has in fact been under construction for at least the past two thousand years in Western cultures.

A fear of the body, aversion to nature, a desire for salvation and transcendence of the earthly plane has created a need for cyberspace. The wealth of the land almost plundered, the air dense with waste, the water sick with poisons: there has to be somewhere else to go.

If you could imagine human consciousness as a fractal geometry, then trajectories of western consciousness would spiral and billow. Here, Eden, Plato, calculus, Descartes, technocracies, and cyberspace are very much part of the "fragmented fractures, self-similar shapes of fractional dimension" of western consciousness.[4]

What if aboriginal consciousness was fractalized, would cyberspace as articulated be part of our geometry of philosophy? Would we have imagined cyberspace? Would we have created cyberspace? I think not—not if cyberspace is a place to escape the earthly plane and the mess of humanity. And not if it is a place of loss or death, where we are to be reborn inside the machine.

Are there points of potential convergence between these world views in cyberspace? The aboriginal idea that all "elements of being, spirit and flesh, sky and earth, can be inhabited simultaneously without active differentiation"[5] may have parallels in the experience of cyberspace. However, the aboriginal view expresses how all life is interconnected; there is no disconnection from the material world. The transformation that is a regular experience in native narratives is not like the experience of escapism in western narrative nor the disembodiment of cyberspace. When Wesucechak changes a young man into his grandfather, he does not become his grandfather. He simply takes on the memories and knowledge of his grandfather, while remaining his grandson. It is a world of subjects to subjects, consciousness to consciousness.

While it may be seductive to draw parallels between aboriginal concepts of transformation, or shape-shifting, and disembodiment in cyberspace, to do so without a philosophical shift would be simplistic. Perhaps some insight can be found in the relationship of animals and humans. A hunter does not get animals because he is a good shot. You could say animals choose to give themselves to the hunter as part of an "old agreement," a symbiotic relationship in which animals and humans communicate. In this view, shape shifting does not occur simply for the thrill of a new body. Instead, shape shifting has purpose—very often a healing purpose.

Cyberspace is here, though, as are native Indian people. In the to and fro of these times, the question is, Can native world views—native life—find a place in cyberspace? Although native world views cannot be easily typified, it is fair to recognize that they embody the desire for harmony, balance, and unity because the universe is viewed and experienced as a place of harmony, balance, and unity. Within this world view, the individual is endowed with the freedom to express and experience singular emotions and thoughts, which are then shared with the community through narrative, ceremony, and ritual—and this reflects the dynamic nature of the universe, of creation as harmonious, balanced, and whole.

How do these concepts fit into cyberspace when cyberspace has been created within societies that view creation and the universe so differently—one that creates hierarchies of being that reinforce separation and alienation with one that seeks harmony and balance with the self and the universe?

At 4Cyberconf, in Banff, during one of the panel sessions, Leroy Littlebear, a professor of native studies and a lawyer, said to the delegates that they should stop this pursuit of cyberspace, if only for a while. In fact he went further and said, "Go talk to trees and to rocks." He also talked about cyberspace as a metaphor and of the West's seeming need to create metaphors instead of emphasizing life.

I imagined him speaking to Columbus, or the conquistadors, or Champlain, or Ponce De Leon, wondering out loud, "What is your journey? What will you do when you get there?"

These words "go talk to trees and rocks" were seized upon. Some in the audience seemed to sigh, perhaps wanting Leroy to be their prophet; maybe they longed for some peace and quiet in the forest. Others seemed to resist and some snickered, perhaps thinking Leroy was referring to the Indian version of "stop and smell the roses."

These reactions, not atypical, seem to offer little thought about the consequences of the journey or the responsibilities in this new territory. Instead, these reactions seem to speak of desire and need for more, always more.

Western culture seems to want everything, to go everywhere. Wants that seem endless, like the hunger of a baby bird, a featherless little dinosaur with its beak open so wide you can see right into its stomach. The desire to know, seek new experience, take new journeys, create light, has somehow grown from a flame to a forest fire that burns everything in its way.

"Talk to trees and rocks" seems too simple for such a hunger. What will cease the hunger, the appetite, if only for a moment?

Of course, in a world with a legacy of colonialism, the hunger of Western culture is threatening and frightening. We have had to feed that hunger, with the furs of animals and flesh of fish and the gold and silver of our lands and ourselves as fearsome mysteries in the West's drama of itself.

In cyberspace, that appetite could well consume "the native," and it has already begun. Terrence McKenna and the advocates of the cybershamanism would take the imagined mind, the supposed dreams of the native, and discard the body—the reality of our lives and the meaning of our shamans. I am no shaman nor their spokesperson but I have been taught that a shaman seeks to restore harmony within the individual, community, and universe. Cybershamanism seems to have developed messianic undercurrents that seem to reflect its advocates' desires more than an understanding of shamanism. Shamans are humanitarians, I have learned, here to gain knowledge and protect the people and the land, not to transcend the earthly plane.

I could continue with how western culture's hunger has damaged the world for everyone. And it is not only the hunger of the West, but the hunger of imperial, colonial, patriarchal orders globally. The philosophy of Indian country, however, is not to be used for a witch hunt. In the imagining of cyberspace, perhaps there are points where aboriginal philosophy can have influence. Tribal cultures are not all oedipal plenitude, we cannot make cyberspace purer for the white folk, but we can concern ourselves with the effect it will have on "the clean land."

One of the basic tenets of our philosophies is a concern for future generations. Each generation considers the consequences of its actions, of its presence not just on the next generation, but often the generation five or seven times hence.

What if with each technological advancement the question of its effect on the seventh generation was considered? Some would say this would limit development. On the other hand, could not such a question free us from the technologically deterministic models that seem to be so persistent today? Neither capitalism nor Marxism feel the need to ask questions about future generations. (Well, they do ask the questions, but the answers are essentially utopian.)

Consider the development of the car, for instance. What of the availability of fuel? What of the consequences of road building? What of? What of? Of course western culture knew little about the environment when the car was invented. The point is precisely to ask the questions about possible scenarios beyond the immediate gains.

Cyberspace will not simply create a new machine, it will fuse human and machine. What then of the world it will create for seven generations hence? Will there be no need for grandparents? Will travel become minimized so that roads fall into disuse? Will paranoia increase? Will there be the need for more or less fluoride in the water? Will we ever see the moon again?

Who considers the seventh generation when creating spaces and narratives in cyberspace? What about virtual museums, virtual malls? Why are these institutions recreated in cyberspace, when cyberspace is supposedly heralding a new world of anarchy or democracy (depending on the age of the heralder)? What will be the consequence for future generations of internalizing the power relation that museums, malls, and arcades imply? What change will occur to their sense of freedom, their free will, as the power relationship between those who own information and those who access it remains much as it does today or perhaps even regresses to feudal ways?

Information as experienced in space and time in a place beyond space and time will no doubt change consciousness. But to what end? Information will no longer be traded within invisible structures of power, but rendered in digital virtual worlds. What kind of information? For what purpose? How will class structure and questions of equity be addressed? As we speak, will our e-mail addresses place us in groups of elites or in groups of plebes?

Much of the future of cyberspace is unknown, and the mystery of its future is fetishized. Questions are indeed asked, but the long term is eschewed in the interest of immediate rewards—corporate, scientific, political, or psychological. And these are not just questions for so-called futurists to ask, but for the makers of fiber optics, for CEOs of cable companies, for computer scientists, for artists, for teachers.

There is an attitude shared by some that somehow cyberspace will replicate tribal society. In this view, cyberspace by virtue of its nature will restore spiritual elements absent from western life, cause the breakdown of central authorities, and renew communal values.

If this was the case, then concern for the future generation would become standard. But so-called tribal behaviors that do not incorporate the respect for consequences have no meaning. I have talked to too many Anglo-Canadians whose idea of "tribal" concerns freedom from consequence. (The very concept of tribal has to be interrogated—that which we as native people enact and that which western culture imagines.)

When native people say that the past is in the future, that the future is knowable, it is not some mystical rant but is rather the result of planning and consideration. As well, it draws on an understanding of the interconnected nature of all life. Prophetic belief and skill aside, there is a relationship to time and space that is not restricted to the moment at hand.

Consider Lawrence Paul Yuxweluptun's virtual reality work *Inherent Rights, Vision Rights*. Lawrence is an internationally recognized artist whose work has been exhibited throughout the world. In his paintings and new media work he draws equally from the surrealist movement and landscape tradition and from Northwest Coast art practices. He creates canvases that tell stories of the genocide of Indian people and the destruction of the land—often with humor. While working in a contemporary art milieu, he uses stories and images to

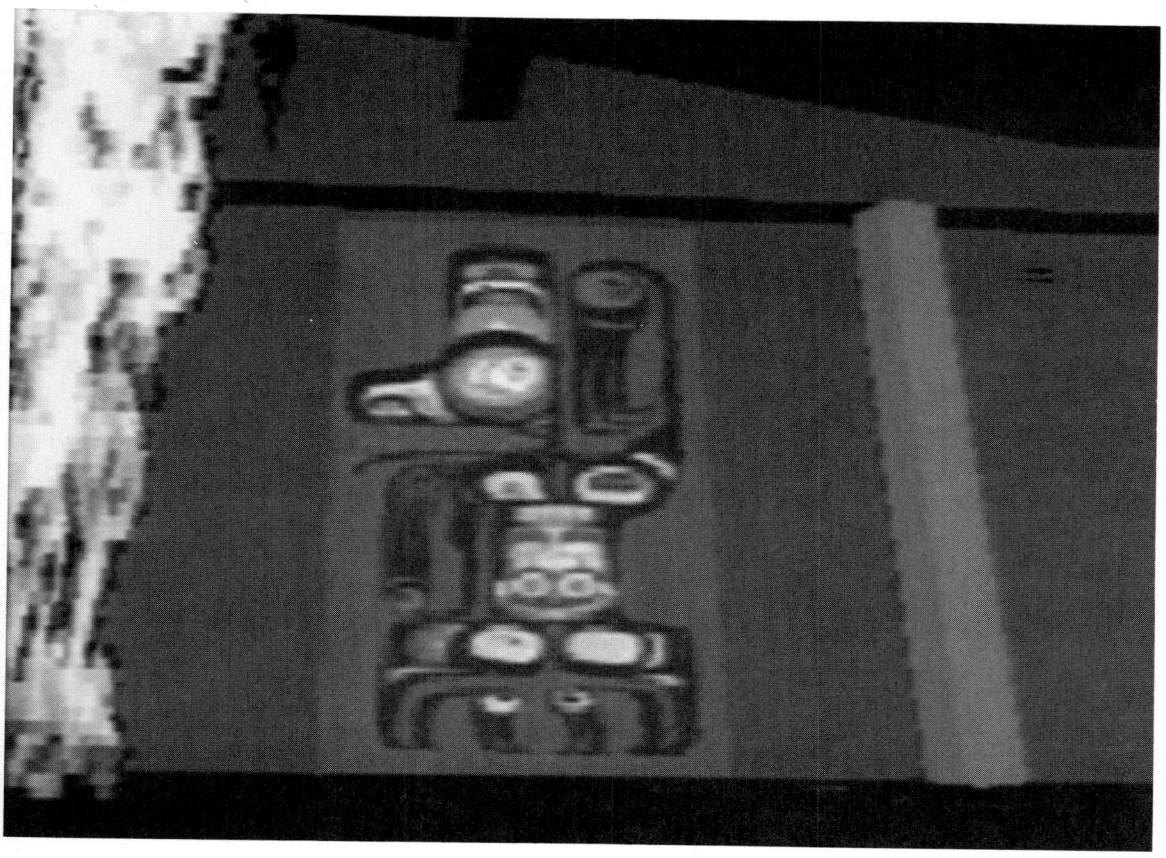

25. Lawrence Paul Yuxweluptun, *Inherent Rights, Vision Rights* (1992), frame grab of the bear inside the longhouse.

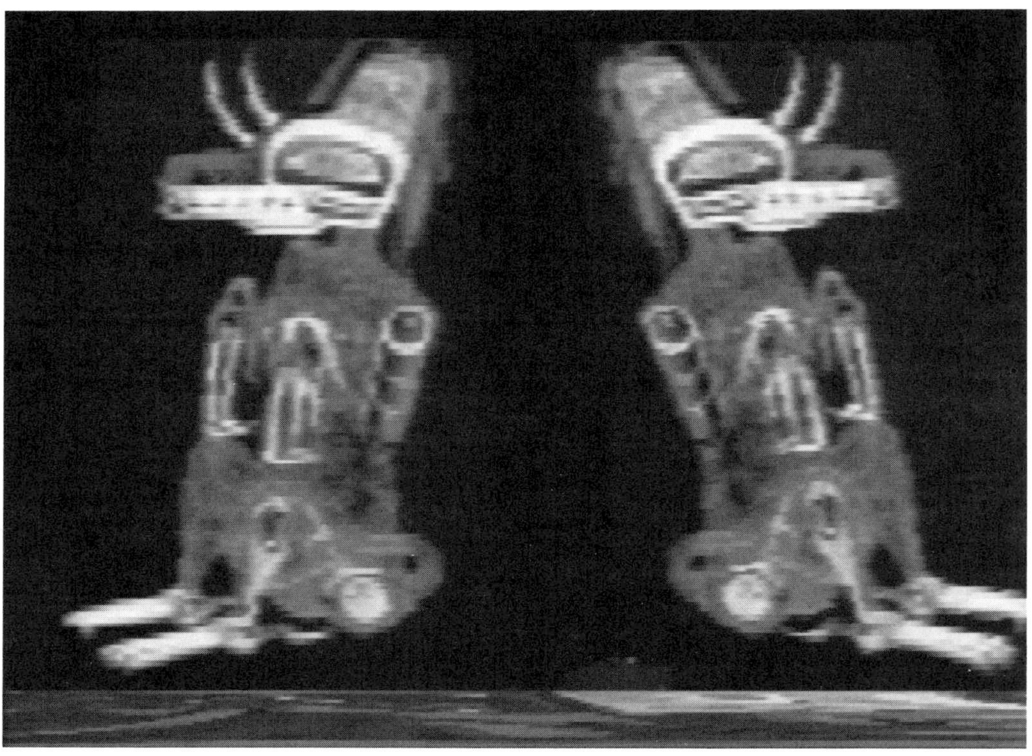

26. Lawrence Paul Yuxweluptun, *Inherent Rights, Vision Rights* (1992), frame grab of the dogs standing guard at the back of the longhouse.

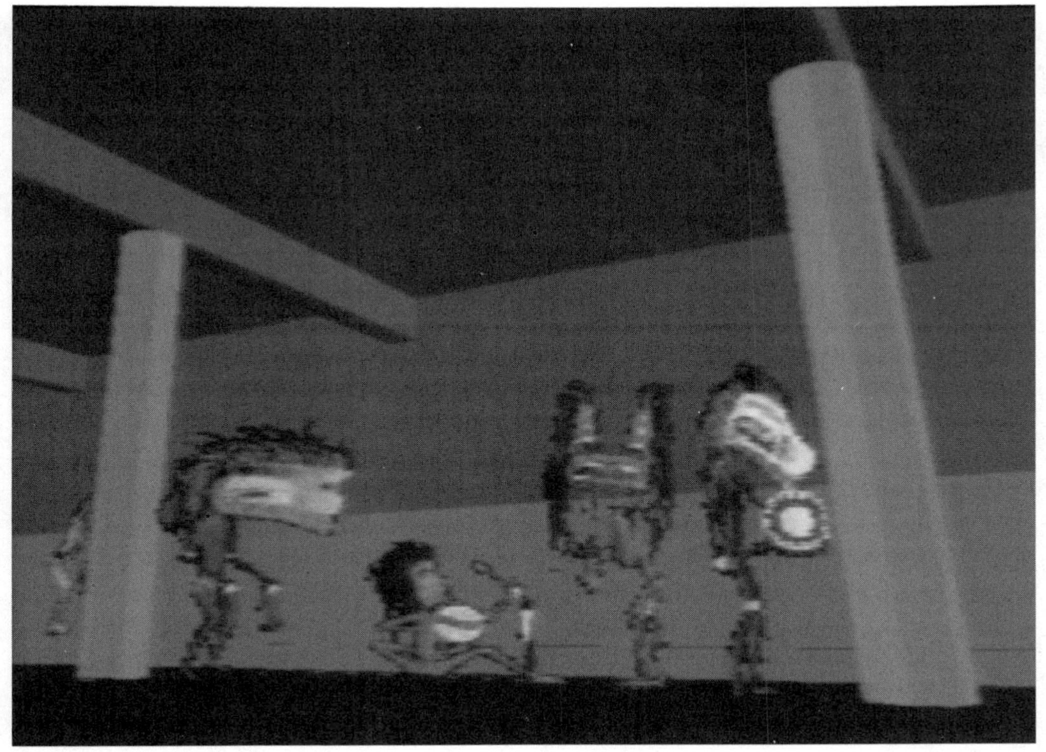

27. Lawrence Paul Yuxweluptun, *Inherent Rights, Vision Rights* (1992), frame grab of dancers in the longhouse.

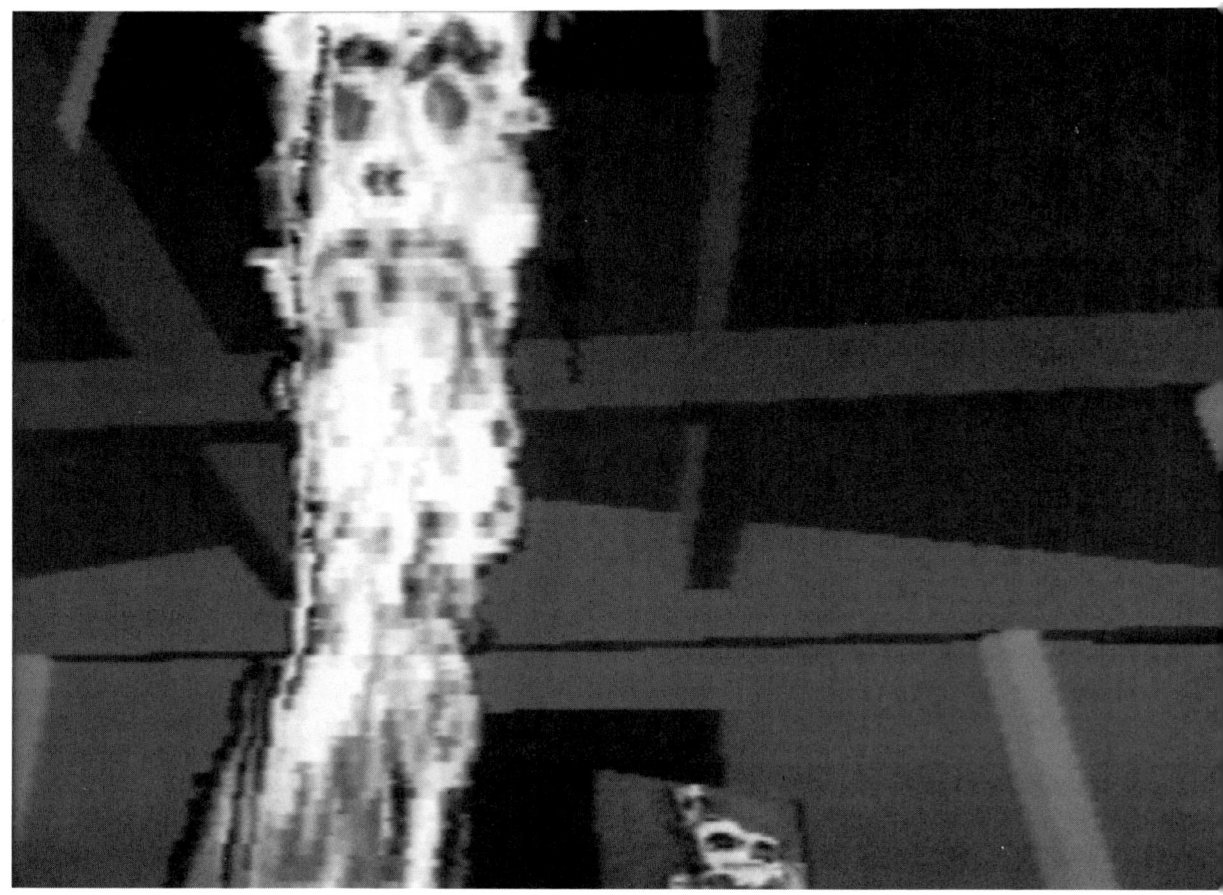

28. Lawrence Paul Yuxweluptun, *Inherent Rights, Vision Rights* (1992), frame grab showing the face in the fire.

demarcate space and time just as they were and still are used in traditional art forms. The story and the history are written on the land.

In Yuxweluptun's virtual environment project there is no helmet, no way to seemingly escape the space you occupy. Instead a kiosk with a viewer, like an old-fashioned stereoscope, allows entry into the work. Once inside, the viewer experiences time. The time of walking from the outside to the inside. The time of listening to a dog. The time of hearing the roar of the fire as the wood burns. Time passes. And there is space. There is a door to enter, to define the outside of the longhouse from the inside. There are four walls and a roof. The smoke rises upward to exit out the smokehole.

Of course, we are in a virtual environment, and the space and time are no more then 1s and 0s organized in front of our eyes, on the screen, and the fire never burns out.

In the majority of virtual environments that I have experienced there is a suspension of time. You move through a seamless-experience corridor with little to mark the beginning or the end. This may serve to undermine the metanarratives of society, no doubt. Here you can experience space without limit. But, to paraphrase, if poetry requires the willing suspension of disbelief, then here virtual reality/cyberspace requires the "willing suspension of the flesh."[6]

Yuxweluptun does not want you to forget your body. Your identity is as intact here as it might be in the material world. You hear your feet along the small pebbles on the ground and will see them should you look down, since you are not immersed in a helmet. Even as you glimpse how Yuxweluptun prays, in the longhouse with the spirit world present, you do not become Yuxweluptun nor a persona he has created through narrative. You are yourself, and must own your feelings and your experience.

In this world a different concept of narrative incorporates agency, which could influence ideas about cyberspace. Here, narrative is derived from storytelling and oral tradition. Storytelling is not exclusive to aboriginal cultures; it is shared around the world, from Herodotus to Scheherazade. Theorist John

Berger explains that "very few stories are narrated, whether to idealize or to condemn; rather they testify to the always slightly surprising range of the possible. Although conceived with everyday events, they are mystery stories."[7] Walter Benjamin compared storytelling with the novel and concluded that in storytelling, "the most extraordinary things are related with the greatest accuracy, but the psychological connections of the events are not forced on to the listener. . . . It is left up to him to interpret things the way he understands them."[8]

Yuxweluptun invites you into the mystery of the everyday. You must define yourself in relationship to the story around you. You are not witness to an ethnographic containment where in the gathering of data you become a hero. Instead you are challenged to "portray yourself," to make your own "psychological connections," since you are your own author.

One could say the experience of cyberspace offers the reversal of narrative as derived from storytelling, a return to an oral tradition, albeit created by text on an electronic screen (for now).

But the parallel would be too easy. It seems to be that despite the subversions of gender transgression, of the experience of nonlinear narrative, of fragmented identities, cyberspace and virtual reality are still anchored to reenactments of western cultural consciousness.

These narratives in cyberspace and VR repeat the traditions of western narrative. Each one experiences their story as a hero asserting his will over others.

In western tradition the hero is like a hunter who gets his prey because he is the best shot or has the best gun or maybe just because he is the hungriest.

In native cultures the hero is like a hunter or gatherer who is only one part of a series of transactions, a process of communication and exchanges between human and animal and the universe.

In scientific terms, maybe the difference is like viewing the universe in classical Newtonian physics of local and mechanistic concepts or in newer concepts that incorporate "non-locality, wholeness and enfoldedness."[9]

In the potential enormity of cyberspace and the seeming limitlessness of VR, the universe seems more "mechanical and separate" than "connected and immanent." The alienated psyche of western man and woman cannot find relief in cyberspace and virtual reality. You can go anywhere, be anyone-but you are still alone.

As science changes, perhaps the universe will be imagined in ways that reflect our interconnectedness. As cyberspace develops, perhaps it will examine augmented versus immersive technology. Perhaps it will explore narrative forms in which you do not leave your body or soul. Just as the storyteller doesn't control the psychological connections of the listener, just as the shaman doesn't invade your mind, perhaps we can create new narratives where you must call upon your own powers and your own words.

There may be other ways to imagine cyberspace, not as a place born of greed, fear, and hunger but instead a place of nourishment. A place where people can find their own dreams. Not just fantasies of abandon, but dreams of humanity and of ways to keep the land clean.

Notes

1. Harold Cardinal, "A Canada What the Hell Is It All About?" in *The Rebirth of Canada's Indians* (Edmonton: Hurtig, 1977).

2. Michael Heim, "The Erotic Ontology of Cyberspace," in *Cyberspace: First Steps,* ed. Michael Benedikt (Cambridge: MIT Press, 1991).

3. Ibid.

4. John Allan Paulos, *Beyond Numeracy* (London: Viking, 1991).

5. Julian Rice, "How the Bird That Speak Lakota Learned a Name," in *Recovering the Word,* ed. Brian Swann and Arnold Krupat (Berkeley: University of California Press, 1987).

6. John F. Friesen, unpublished essay, Vancouver, 1994.

7. John Berger, *The Sense of Sight* (New York: Pantheon, 1985),

8. Walter Benjamin, "The Storyteller," in *Illuminations: Walter Benjamin,* ed. Hannah Arendt (New York: Schocken Books, 1969),

9. F. David Peat, *Einstein's Moon* (Chicago: Contemporary Books, 1990).

Nature Morte: Landscape and Narrative in Virtual Environments

Margaret Morse

> The phantom worlds and pseudo worlds of cyberspace could also destroy (have also destroyed) the quiet inner narrative on which the nineteenth and twentieth century self was built.[1]

Introduction

At one time, travel through a landscape was both a practice and a metaphor that served to constitute the self: "Subjective identities themselves were grounded in narrational journeys."[2] But the very notions of landscape and narrative, not to mention the activity of travel, have become problematic in cyberspace. To explain why, this essay describes certain aesthetic features of cyberspace and virtual environments. It also suggests why art that is engaged with information society is not limited to the spheres having access to information technologies. Then it addresses two recent experimental virtual environments produced at the Banff Centre for the Arts in 1993: Michael Naimark's *See Banff!* and Toni Dove and Michael Mackenzie's *Archeology of a Mother Tongue*. Just how "subjective identity" may change in the process of constructing late twentieth- and twenty-first-century electronic culture remains open to speculation, yet the very act of producing virtual environment artworks invites us to step inside to find out.

Propositions about Art and Cyberspace

Whether we call them *virtual environments* or *cyberspace,* the gathering places immersive computer-graphic worlds—are increasingly situated in what amounts to *non-space*. Unlike the physical world, which is always "full" and ready-made, the primordial virtual landscape (as pictorially described) is an utterly empty computer-generated display that, to date, is stocked sparsely and is largely constituted from scratch with considerable graphic effort. Most virtual environments remain relatively void but for a crude symbolscape of geometric objects taking on a variety of metaphoric shapes. Once these graphics are lost from sight, it is easy to get lost in the unboundedness of cyberspace—a uniform, three-dimensional, and possibly infinite void.

So far at least, there is no direct or transparent technical continuity between the cultural forms of the mechanical age and the developing forms of the information age.[3] So it is not as easy as one might think to simply transfer intact the immense cultural capital of image culture into virtual fields. How, for example, does one translate the vista of a specific site into the representation of it as a framed landscape in the form of two-dimensional, photographic (that is, analogical and indexical) realism and then again into a three-dimensional, digitally produced, utterly symbolic virtual environment? What could mediate this process?

But there is a prior question one might rightly ask, as did aboriginal participants during 4Cyberconf at the Banff Centre for the Arts: why the allure, what is the attraction of cyberspace? Why produce virtual environments at all when we could be enjoying a far more satisfying and beautiful physical reality? The latter question is similar to that implicit in Raymond Williams's *Television: Technology and Cultural Form*[4] written thirty years earlier: why television? For Williams, television became a mechanism for cultural integration through its social control and communication functions. Its need was created by an industrial economy that uprooted much of the population, divided work from home, and isolated one person from another in privatized forms of living, such as the separate dwellings of suburbia. Highways may link home to work and commerce, but they do not overcome the isolation of "mobile privatization." Television broadcasting, on the other hand, offers culturally unifying experiences and can even encourage relations to programs that substitute for some aspects

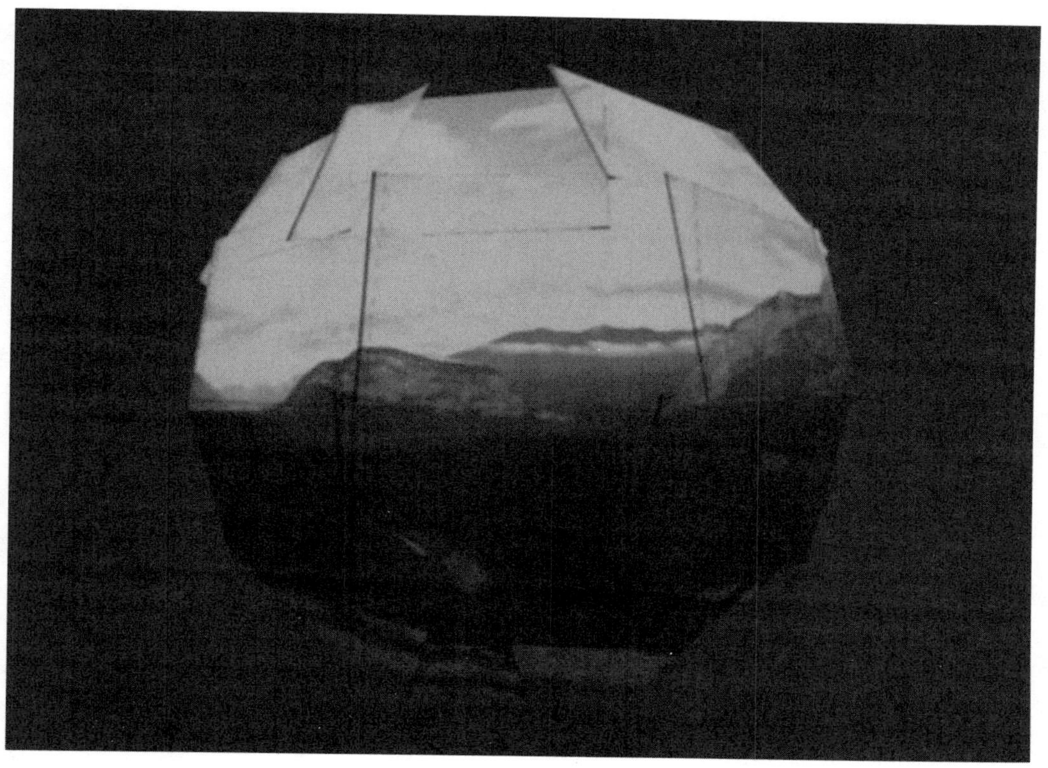

29. Michael Naimark, *Study #1* (1992). Digital image John Harrison.

of human interaction. Then, the allure of television has deep roots in the need for human contact, identity maintenance, and a sense of belonging to a shared culture. The computer-based electronic networks into which television itself is being integrated serve the far greater and more complex needs of a postindustrial and postnational sociopolitical information economy. This economy is now the excuse or the occasion for a wrenching restructuring of the workforce that displaces some people and brings others together electronically—but all are separated from each other in physical space.

The electronic sphere of information is still very much in the developmental process. However, even those excluded from this virtual realm will likely suffer the uncertain effects on economies and cultures throughout the globe. Furthermore, there is an implicit social hierarchy determined by access into this virtual realm, which divides have and have-not far more drastically than mere walls. Can First Nations avoid encounter with this larger process? Probably not; the issue is not then whether cyberspace, but which cyberspaces, for whom, and how. It is then a task of art to give shape to possibilities and questions about this socioeconomic shift and its emerging cultural forms that are not raised by virtual environments produced and designed for instrumental purposes or for entertainment.

Once "inside" cyberspace, what happens to subjectivity traveling in the "nonspace" of a virtual environment? With a head-mounted display (or "eye-phones") and clothing (data suit and/or glove) the user need not leave the spot to be *there*. In effect, while the visitor to the virtual environment moves in a very circumscribed physical area, his or her motion is tracked and the appropriate shift in his or her point of view within a vast virtual landscape is constructed instantly. Cyberspace, then, is not merely a scenic space where things could happen; it also incorporates the artificial intelligence or *agency* that orchestrates the virtual scene (delegated human subjectivity). This intelligence can even be invested with aspects of personality, challenging the rational distinction between space versus agency or landscape versus character in the process. On the other hand, animistic, folkloric, or poetic attributions of agency to the earth and to features of the landscape are given now a virtual sphere of reference.

When the "reader" or receiver is virtually inside the scene or story-world in cyberspace, enjoying choice-making privileges about the order, direction, or pace of what happens next, he or she enjoys a role somewhere between that of a character and a narrator in the virtual environment. Surrogates of the user within the virtual realm can be expressed in many different persons and degrees of immersion: an "I" or the subjective and "embodied" view of the world from inside it; a "me" as a corporeally separate persona or avatar, whose appearance and characteristics (often chosen from stock) represent the self in a screen-based world; a self that lurks as a ghostly, disembodied perception, marked or unmarked in that world; or a character, "he, she, or it," with a more distanced relation to the visitor's self—and there is the uncanny agency of the space itself. Furthermore, the voice of a controller/programmer/author or coparticipant may leak in from "outside" or the view of the virtual world may be superimposed over physical space.

With these options, how can just one story "quietly" unfold in the linear and coherent way stories were supposed to, now that the reader or viewer becomes a *visitor* or a surrogate of him or herself (as opposed to identifying with a character) inside the story? Some speculate that a subject immersed in virtual reality, with its mobile perspective and multiple narrative paths, would lose its identity, splinter, and fall apart.[5] Perhaps the opposite is true—the multiple aspects of "personhood" and "agency" in the landscape of cyberspace, like the "second self" in the computer, are part of a long-term cultural trend (of which television is the greatest exponent so far) in which more and more of the task of enculturation that produces subjects out of human beings has been delegated to machines. Then, there may be more continuity between the experience of the physical, the electronic, and the virtual environment than the fragmentation thesis accounts; the many aspects of person available to the user of a virtual environment may offer more possibilities for subjective integration and control, not less. The cybersubject, no longer socially determined by a physical appearance, is free to proliferate personalities—but is still linked, for now, to one physical living body. What subjectivity will become in information societies is still an open question. Meanwhile, that material body, tethered to a tracking device, much like Ulysses bound to the mast, hears the Sirens "inside" on display, while the oarsmen made deaf to their song continue rowing.

When asked what cyberspace is today, William Gibson, the science-fiction author who invented the term, answers, "It's where the bank keeps your money."[6] Cyberspace serves and is shaped by the needs of postnational and postindustrial corporations: knowledge decontextualized and digitized as information is stored in databases and manipulated. Whatever the interests and applications involved, however, for the vast majority of users information must be displayed as symbolic forms and made perceptible—cyberspace proper, so to speak.[7] However, the forms of cyberspace are not solely virtual: material objects and environments that have been invested with computer-supported agency (that is, delegated human subjectivity) or that have been given the skins of other times or places are also cyberspace. In fact, for an economy based on virtual money to work, as in the artist Jeff Schulz's model of the credit system,[8] the system of exchange value seems to demand passage from the conceptual to the virtual to the material and back again, crossing through a variety of reality statuses.[9] In the same way, electronic and/or virtual environment art is linked to other artistic mediums by metaphors across different degrees of materiality.

While cyberspace per se is an exclusive realm, its production depends on the material space beyond its interfaces. Yet even those who assemble the computer chips may not have access to the worlds that computers engender. Since electronic networks involve a choice about who will be connected and who will not, a network consists of its gaps as well as its links and nodes. There is a negative or shadow cyberspace—material and devoid of technological resources—that those who seek to understand electronic culture must take into account. That is why those who are technologically barred from it are nonetheless discursively and actually a negative part of that system; they can offer valuable and enlightening views about cyberspace in what must be a truly global dialogue with local positions.

Cyberspace, then, is the manifestation of what some call the datasphere in perceptible—and that means largely metaphoric and mostly ubiquitous and unavoidable—forms. Perceptions of cyberspace as nightmare and/or utopia are understandably related to one's position in this economy of virtuality and the mode of access to it, if any: the data entry worker is in a different position than the programmer, a little girl has a different cultural entitlement to cyberplay than a little boy; the subsistence farmer's life, if not status, could not be more different from the fast food worker's, but they will nonetheless be ultimately related in a global system of integration and exclusion, like the strands and negative space of a net. Questions about the transitions from material to virtual, ranging from representation of landscape, politics of access, agency, subjectivity and narrative, all come to the surface in new virtual environment artworks. How can we approach them?

Three Axes of Cyberspace

There are at least three axes along which one can situate cyberart.[10] The first axis is the exploration of the mixtures and links between virtuality and materiality to be found in the everyday life of information societies. The world full of remote controls and "smart" appliances, automobiles, computers, and video cassette recorders is an enhanced or augmented reality of distributed cyberspace in which some of us already live. We interact daily with acting, decision-making, writing, and speaking machines informed by a store of data accessed in "real," that is, unreal time. But a cyberized physical reality is even to be found in the pockets and zones of the natural world that have become commodified landscapes, such as touristic Banff. No longer just material trees and rocks and waterfalls with their own life, invested with divine intentionality, Banff's tourist sites are veiled with the varnish of postcard images, symbols of themselves in touristic exchange. (Perhaps we could even include street lighting in this category as proto-cyberized reality, as it changes night into something artificially in between night and day.)[11] The point of such image technology, including the technology of cyberspace, is then far from neutral: it is, in the words of N. Katherine Hayles, "to reconstitute the body," and, I would add, the landscape, "as a technical object under human control."[12] Art that foregrounds the links between the material and the virtual can question this domination; it can take forms in many different media and genres, including robotic and screen-based interactive pieces, which must ultimately be a self-interrogation by the virtual image of itself.[13]

The second axis of cyberspace includes all those linking devices that create virtual spaces of greater and greater, albeit ephemeral, unities—text-based networks, Multi-User Dungeons (MUDs) and Multi-Object Oriented networks (MOOs), telecommunication satellite links and cables, and also proto-cyberspace devices such as the railroad and highway networks that unify physical space. This realm of connections invites the study of transitions between spaces and conditions, as well as the analysis of crowd movements in and between worlds—including the popularity of a topic on the net, its ebb and flow. The attractors and repulsers of exchange have gained the attention of theorists and artists. Such investigations finds their inspiration in the representation of the crowds or group protagonists in, for instance, films by Eisenstein, Buñuel, Jansco, Akerman, and Altman, in the satellite experimentation of early video, and the use of time-lapse studies, flocking algorithms, and the statistical flows and ruptures of chaos theory for the purposes of art. Then, the vocation of an art that is of and/or reflects on such exchanges is not the representation of the visible world, but the visualization of an entity or flow that is otherwise imperceptible because of its cosmic or microscopic scale, its discontinuities in physical space, or its intervals in time.

The third axis of cyberspace is the solely virtual or immaterial world with which one interacts to varying degrees and within which one can enjoy varying modes of person. The entirely virtual world is the most traditional kind of cyberspace and a last gasp of Renaissance space, in that the spectator's station point is inside the projection of an image. In fact, the visitor's look itself appears to be creating (or performing) a world that is actually activated in "real time" by a computer from a digital store by tracking the visitor's every move. While virtual worlds can seek to reproduce natural landscapes, there is a thrill of transgression in entering a symbolic landscape or metaphor of what would otherwise be inaccessible and impenetrable—such as the inside of the body, the atom, the black box, the outsides of our galaxy and our universe, or other worlds entirely that have no counterpart except in fantasies and dreams. For the most part, to be inducted into a three-dimensional virtual environment is to go where people cannot go but computers can, into a landscape that is nano-, macro-, or otherwise out of human scale or that would otherwise be invisible beyond opaque surfaces of material or skin. To enter the virtual environment itself is like being able to walk through one's television or computer, through the vanishing point or vortex and into a three-dimensional field of symbols. It is as if one were immersed in language itself or as if the symbols on a map were virtually embodied as landscape. The virtual landscape in question might be imaginary because it never existed or never could exist or once existed and is now lost, lost, lost. (Salvage anthropology which seeks to preserve disappearing behaviors and decaying or dispersed artifacts has found its appropriate medium.)

Any virtual environment, as instrument or as art, can be analyzed by means of all three referential axes: first, the relation of cyberspace to the external world, be the world "natural," urban, or the realm of fantasy and dreams; second, the internal links and relationships within cyberspace; and, finally, the symbolic relations within any virtual environment. In specific environments, however, one axis may be thematized or drawn into the foreground. Two of the artistic experiments produced at Banff help demonstrate how virtual environments can be understood by these axes: Michael Naimark's pieces emphasize the relationship of the cyberlandscapes and the material landscape, while Toni Dove and Michael Mackenzie's work foregrounds the dreamlike condensation of many different reference points into one symbolic landscape. Beyond a discussion of the production of personas or characters within virtual worlds, work on the axis of networks and links between worlds in cyberspace was not represented to a significant extent at 4Cyberconf or the Symposium on Art and Virtual Environments. Perhaps the designation "virtual environments" in the title of the project and the symposium unintentionally promoted such neglect. The idea of networks and links as "spaces" or environments is somewhat of an enigma, still to be forged and integrated into thought, rather than an accepted notion.[14]

NATURE MORTE: CYBERLANDSCAPES AND DEATH

Nature morte as it is used in this essay no longer refers to the genre of still-life painting, but to an ambiguous relationship between an animate virtual environment, the mortal body, and a fading natural world. While the aim of instrumental uses of cyberspace may be smooth fit and resemblance between the virtual and the real, the art of virtual environments may prefer to foreground the shimmering inexactitude between the material and the immaterial, and to allow for ambiguity in the apparent association of the virtual with the seemingly immortal, infinite, and sublime. In his treatment of the features of postmodernist fiction in literature, Brian McHale identifies "opalescence" with uncertain ontological status, a place where worlds overlap.[15] Perhaps the virtual landscape is all the more uncanny as it is not precisely fiction, but an immaterial state that can have real effects, just as the virtual person is an ambiguous being, neither completely alive nor dead and not entirely here nor there. Perhaps, one task of art in an information society is to explore the threshold condition of animate undeath that is the virtual itself in its contemporary development.

The quality of "opalescence," or what Robert Kyr called "shimmering" sound when describing the mix of gamelan tunings with electronic music, is prominent in many examples of cyberspace as art. Consider the wavering path created when Laura Kurgan used global positioning satellites to mark the location of her installation at the New Museum in New York, suggesting the play of the oscillating satellites above the earth. This trembling quality is visible in Toni Dove's pulsating and glowing images of girls in Dove and Mackenzie's *Archeology of a Mother Tongue.* In *See Banff!* Michael Naimark's simulation of a kinetoscope evokes both a period before the cinema and one after; the flicker of noticeable intervals between frames is simulated as well, when the stop-motion stereoscopic images of touristic Banff are "cranked through" the machine. The landscape in both pieces is bathed in the opalescence of worlds not quite coalescing. The spectral, "shimmering" aspect of cyberspace suggestively raises the sense of mortality that infuses these works; just what or whose mortality is ambiguous? Possibly that of an individual or social body or perhaps the larger body of

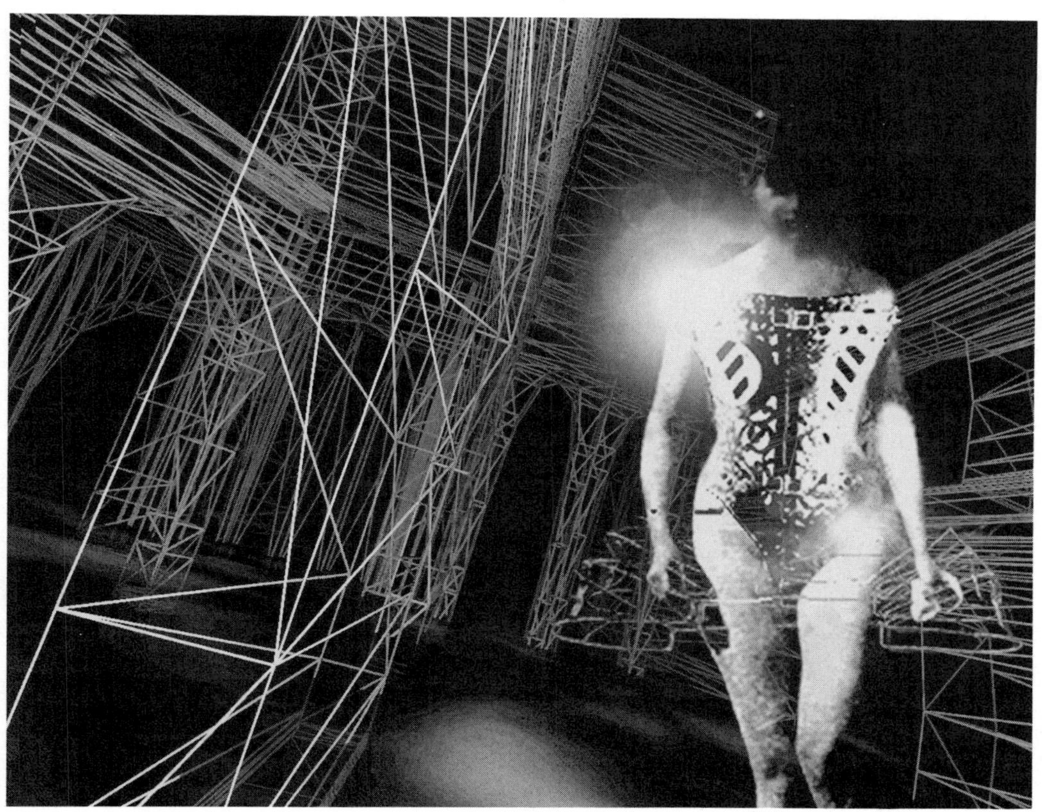

30. Toni Dove and Michael Mackenzie, *Archeology of a Mother Tongue* (1993), frame grab from one of three narrative "envelopes" or virtual environments. Digital image courtesy the artists.

what was once called the "natural" world in a situation of socioeconomic and cultural transformation.

Landscape as Underworld

When discussing cyberspace or an animate virtual landscape, one must consider death as well as life. The best-known cyberpunk novel, *Neuromancer,* is explicit in this connection: the hero, Case, meets with his dead wife toward the conclusion of the novel via a boy. He asks the boy's name:

> "Neuromancer," the boy said, slitting long gray eyes against the rising sun. "The lane to the land of the dead. Where you are, my friend. . . . Neuro, from the nerves, the silver paths. Romancer, necromancer. I call up the dead. But no my friend," and the boy did a little dance, brown feet printing the sand. "I *am* the dead and their land." He laughed. A gull cried. "Stay. If your woman is a ghost, she doesn't know it. Neither will you."[16]

Note that in this excerpt, the term *neuromancer* wavers between reference to a computer-produced spatiotemporal realm within which Case and his dead wife meet, the artificial intelligence that produces that realm, and a persona standing for it and acting within it—not to mention Gibson's book itself.

As a result, cyberspace, beyond its business uses, can invoke a parallel and sometimes a transcendent or spiritual world that revives the dead or the spirits of things in the limbo of the possible. Just as the mysterious telegraph and the telephone were also once thought able to contact the other world, cyberspace is also an underworld in which to meet one's Eurydice.

The Ultimate Display

On the other hand, cyberspace can also be about *control* of the body and material space. This is especially the case when the signs of nearly everything in the material world (except perhaps humanity) are displayed in a virtual world perfectly matched to our own in a point-to-point indexicality: a landscape displayed on a flight control panel during the Gulf War could be as deadly as the virtual maps that guided the trajectory of Cruise missiles, coordinated under the constant gaze of satellites. In this war that supposedly would not kill people, just infrastructure, the aim of manipulating symbols in a virtual landscape was to obliterate the actual referents of undesirable symbols in real space. The worlds of entertainment, while not actually deadly, commonly rely for a sense of finality and realism on the killing function for the game's "finishing moves"—decapitation, tearing out the heart, blowing up, eating, and so on.[17] Death is apparently the guarantee that the joystick controls the image and that the map controls the territory, while "death" is the measure of mastery in the game, just as the material body and its ultimate death is the warrant for all the personas and avatars that lurk and wander through the net.[18]

Ivan Sutherland exposed this logic at its most extreme at the inception of contemporary work in virtual imaging systems nearly thirty years ago with his "ultimate display":

> The ultimate display would, of course, be a room within which the computer can control the existence of matter. A chair displayed in such a room would be good enough to sit in. Handcuffs displayed in such a room would be confining, and a bullet displayed in such a room would be fatal. With appropriate programming such a display could literally be the Wonderland into which Alice walked.[19]

Sutherland's Wonderland is remarkably sparse: a chair and a pair of handcuffs suggests that one is chained to the chair somewhat like a prisoner in Plato's cave. The room also possesses a fatal bullet and, by implication, somewhere inside or outside the room, a body—it is not clear whose. Note that the realism of the ultimate display does not depend on appearing like reality or verisimilitude or *referential illusion* (that is, referring convincingly to another world elsewhere), but on the power of language or the display to make reality itself, or *enunciative illusion* (that is, appears to call a symbol forth into existence by speaking or drawing it; or what is known in linguistics as performative speech). This power of the symbol or the word to give life is paradoxically demonstrated or confirmed only by a performance that puts a body (a real body?) really to death.

Beyond such deadly performatives there are more gentle displays that neither represent nor kill the "real" world as we know it, though they may change it socioeconomically and culturally. Virtual landscapes are liminal spaces, like the cave or sweat lodge (as Lawrence Paul Yuxweluptun's work suggests), if only through their virtuality—neither here nor there, neither imaginary nor real, animate but not living and not dead, a subjunctive realm wherein events happen in effect, but not actually. It might seem that the limits of the body and material existence might be transcended within such a subjunctive realm of play, if only for a little while. Of course, the spiritual potential and powerful metapsychological effects of immersion can easily be abused as a seductive force, even in the more mundane, instrumental uses of cyberspace. Yet the very notion of *immersion* suggests a spiritual realm, an amniotic ocean, where one might be washed in symbols and emerge reborn.

The underworld in which symbols of the once-was, is-not-yet, and never-will-be are given virtual life is poles apart from the ultimate display in which the virtual proves its power over material reality—first by resembling it, then by submitting it to death. These two relations to, or motivations for producing, cyberspaces contrast an open-ended and subjunctive realm of fantasy against the desire to control the material world via a virtual double. While Naimark's and Dove and Mackenzie's pieces produced at Banff emphasize different axes of cyberspace, each of the works addresses these two poles of fantasy and control. Dove and Mackenzie's *Archeology of a Mother Tongue,* presents a dreamlike condensation of many worlds into one, but it also addresses the desire to control the material world and its contingencies as one of its major themes via the figure of a pathologist. Meanwhile, Naimark's *See Banff!* is an implicit critique of tourism and its domination of the material world as *nature morte;* yet the piece is also invested with a sense of the underworld and what escapes the desire for mastery.

SURROGATE TRAVEL: *SEE BANFF!*

> While our sense of the natural world has always been encumbered by our sense of human culture and history, there was a time, not long ago, when you could get out of your car at a curve on a scenic road and admire the view on something resembling its own terms.[20]

The social practice of travel is driven by romantic desire for such transformative symbolic experience in an *other* place, preferably a paradise, from which one could return renewed. Tourism converted secular pilgrimage into a commodity marketed in two-dimensional images; mass tourism reconstituted the three-dimensional landscape itself into a "technical object under human control." The dynamic through which a natural site becomes a theme park of its own ideal image can eventually destroy the attractiveness and aura of authenticity that first drew crowds to it. Put most drastically (and possibly controversially), the touristic sight invites the look that kills, albeit slowly, in a deadly spiral of development in which the virtual has been implicated from the start.[21] Surrogate travel in a cyberlandscape would merely extend this logic by allowing the viewer to virtually enter inside the image. Could a reflection on tourism be staged without reinstigating this deadly process or enticing more footfalls onto fragile paths? The problem for the artist is the perennial one of how to employ the very mechanisms and apparatuses of vision that support the practice on which one wishes to reflect.

According to Alexander Wilson, to view the natural landscape in the mid-twentieth century was to see a world largely blanketed with information. Dean MacCannell's *The Tourist*[22] provides an explanation of how such information in the form of a "marker" or label produces a "sight," in this case, a specific touristic destination around Banff in the Canadian Rockies. In such a way, the picture postcard and its legend become not so much a way of framing and capturing the natural world as a way of producing an image and a set of connotations, which the natural world, when visited, is expected to respectively resemble and evoke. That is, the landscape is no longer blanketed with signs, it *is* a sign, even naked before the eye. As Christopher Pinney explains, in the logic of tourism in western culture, the journey itself frames a foreign or *other scene* for the traveler, who has seen it already constituted as a picture or image before. To traverse the actual referent of the picture in physical space—a "real-space," as Naimark would term it, that nonetheless retains some of the character of an image or map—adds, according to Homi Bhaba, "that dimension of depth that provides the language of identity with its sense of 'reality'; a measure of the 'me' which emerges from the acknowledgment of my inwardness, the depth of my character. . . ."[23] Then we might say that surrogate travel through a virtual environment simply realizes virtually what is already constituted as the goal of the journey: being there *in* the image. It is but one step more in this logic for the sight itself to become a landscape composed entirely of information, a virtual graphic realm sustained by computer.

Michael Naimark has long been involved in projects of "realspace imaging" involving surrogate travel and "moviemaps" using photographic images of a particular landscape: an interface and a screen or monitor allow one to see, for instance, Aspen by "car" (created with the Architecture Machine Group at MIT from 1978 to 1980), San Francisco Bay by "air" (with the San Francisco Exploratorium in 1987), or Karlsruhe by "tram" (with the Zentrum für Kunst und Medientechnologie Karlsruhe in 1990 and 1991). Better than viewing it on screen, the surrogate "driver" or "pilot" is capable of moving at a far greater speed and can move backward as easily as forward on a grid of predetermined paths, doing what Naimark calls "browsing." At Banff, Naimark extended his investigations into the tourism of the natural landscape in three stages: the first field studies in 1992 were an almost casual bricolage of local sites, displaying an unconventional informal attitude in a field conditioned to depend on computers the size of Coke machines. While he hardly eschews leading-edge technology, Naimark's work is characterized by a mixture of old and new apparatuses—such as the use of a baby-rickshaw, slide film, laser disk, and computer technology to produce *See Banff!* In his second experiment at Banff, a model of Johnston Falls produced for Brenda Laurel and Rachel Strickland's *Placeholder,* Naimark transformed mobile video of the falls into a flat, still image

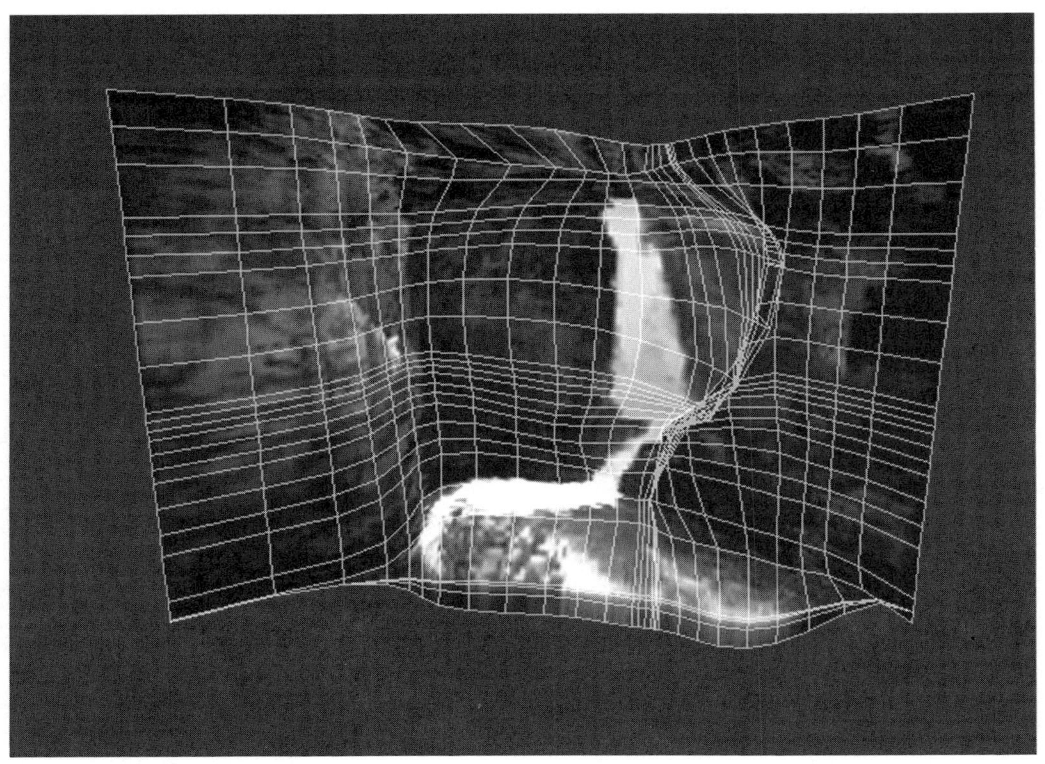

31. Michael Naimark, *Study #2* (1993). Digital image Cathy McGinnis.

floating in the cybervoid and shaped it like "mashed potatoes" into a relief map of itself. With *See Banff!,* his formal experimentation continued but was conditioned by explicit reference to tourism. By presenting the piece in an apparatus that recalls the now century-old kinetoscope and the naïve enthusiasm of the turn-of-the-century stereoscopic card series "Journey of the World," *See Banff!* gathers social and symbolic resonance. In the process, the artist lends surrogate travel a critical and self-reflexive edge. Formal issues addressed in the earlier field studies set the stage for understanding the sociocultural issues around tourism and their treatment in *See Banff!*

Field Studies

Naimark's 1992 demonstration tape, *Panoramic and Moviemap Studies in Computer 3-Space,* is a description of the process through which he sought to meld photographic space and cyberspace. His videotaping forays produced three segments: a panorama of the Banff river valley, a depth or *z*-axis move through a recreational vehicle park, and a traveling shot of a Rocky Mountain vista from a moving car. Then he selected specific frames from each field study and texture-mapped them over one of three different wireframe geometric shapes floating in the cybervoid. Despite being in sculptural shapes, the photographs that compose the virtual world retain their two-dimensionality, appearing more as stacks or fields of photographs than as a convincing landscape. The results are paradoxical, the effect of this incongruent mixture being to undermine the illusion-producing capacity of both the two-dimensional photographic and the three-dimensional immersive modes of spatial representation. In effect, Naimark has solved the problem of the flatness of photographs when introduced into a three-dimensional virtual world by underlining their two-dimensionality.

The river valley panorama, for instance, was taped by swiveling a tripod-mounted video camera systematically like an apple peeler in a circular axis around the landscape. The resulting images then were texture-mapped over a stationary hollow sphere that the visitor to its virtual space could navigate in and around. Seen from outside, the resulting landscape is like a design on a ball; from inside, the sphere is a partially immersive virtual landscape—partial because it also conveys the sense of being a paste-up of flat photographs. The effect might be compared to one of David Hockney's photo collages, except that Hockney's pieces actually *are* paste-ups of photos spread over a flat surface.

The second field study of a *z*-axis move through a campsite is texture-mapped onto a stack of flat images lined up like sheets of photographic paper on a spike, or, in Naimark's description, like "an exploded deck of cards." Virtually "driving" through image after image in quick leaps and jerks is like a subjective view of a skewer assembling a shish-kebab, breaking through card after card. (Such stacked images are actually a basic transitional device on television news and other genres, expanding the notion of "suture" far beyond its reference to film theory.)

The final segment, a traveling shot out the passenger window of an automobile, produced a particularly confusing image for viewers of his video documentation, according to Naimark. The reason why is easy to understand: by laying his selected images end to end in cyberspace like a filmstrip, the "temporally undersampled" trees and other features on the landscape in the foreground pop up and disappear left and right in frame, effectively violating the rules of film continuity. Meanwhile, despite the fact that the foreground images and the vector of motion are traveling to the right, the image of the huge mountain range in the background remains relatively constant. Thus, the image undermines the notion of a stable landscape over which one travels, not to mention the sense of traveling itself as covering ground. It is this study that suggest how paradoxical "travel" can be when movement that makes sense in physical landscape is translated into the disparate imaging technologies of video and the computer. The artist has also demonstrated the relative nature of movement in cyberspace: whether landscape or subject, moving objects can be made still and vice versa.

See Banff!

Michael Naimark's kinetoscope simulation depicting "thereness" in Banff similarly began with field studies. The fifteen sequences (plus credits) selected from over one hundred for *See Banff!* explore different sites, different deployments of the mobile setup, and experiments with timing in mobile setup (which consisted of two synchronized 16 mm film cameras mounted side-by-side, "baby-like," in a jogging carriage). Naimark's computer-driven video disk and stereoscopic display offers virtual journeys in the Canadian Rockies and environs of Banff through a peephole for one individual viewer. However, the piece is not a nostalgic reproduction of scenic postcards come to life, but a condensation of features from different eras in the history of tourism as it has been shaped by the photographic image: the kinetoscope, the stereoscope, the movies, and now surrogate travel or "realspace imaging" in a virtual environment.

The vistas are at once visually hyperreal and temporally unreal in a simulation of the flicker of animation. *See Banff!* does not produce a smooth and coherent world, but rather a flipbook with bits of virtual space-in-between. Different deployments of the camera reveal different facets of the sight: the stationary camera reveals the spectral motion of the tourist figures, while the laterally tracking shot reveals the normally repressed or backstage landscape, be it the devastated spillway of the Oldman Dam or the huge geometric forms of Albertan grain silos. However, the most typical move in the series is the *z*-axis or motion into the image, for instance along a trail to Johnston Falls. Oddly enough and by Naimark's choice, the *z*-axis reveals very little of the landscape and much more of the trail and the traffic on it. The viewer is implicated in the space of the virtual image not only through stereoscopic perspective and as voyeur or nature junkie, peeping at postcards of the Canadian Rockies, but also as a surrogate traveler. *In* the image, the viewer cranks the pace of his or her own motion faster or slower, backward or forward at will down an obligatory trail hardwired not only into Naimark's program but into the earth. Repeating a sequence over and over or moving the crank back and forth produces rapid and surreal appearances and disappearances of swarms of tourists in Brownian motion: it does not support a sense of subjective experience, rather it conveys the effect of an anonymous mass in a ritual with strict itineraries. Even the spectacular vista of Lake Louise must be glimpsed through the real subject of the segment, which includes the displaced urban bustle of tourists in the picture, ordinarily taboo for postcard images because it makes them no longer pure and "just for me," thus diminishing their picturesque or sublime character.

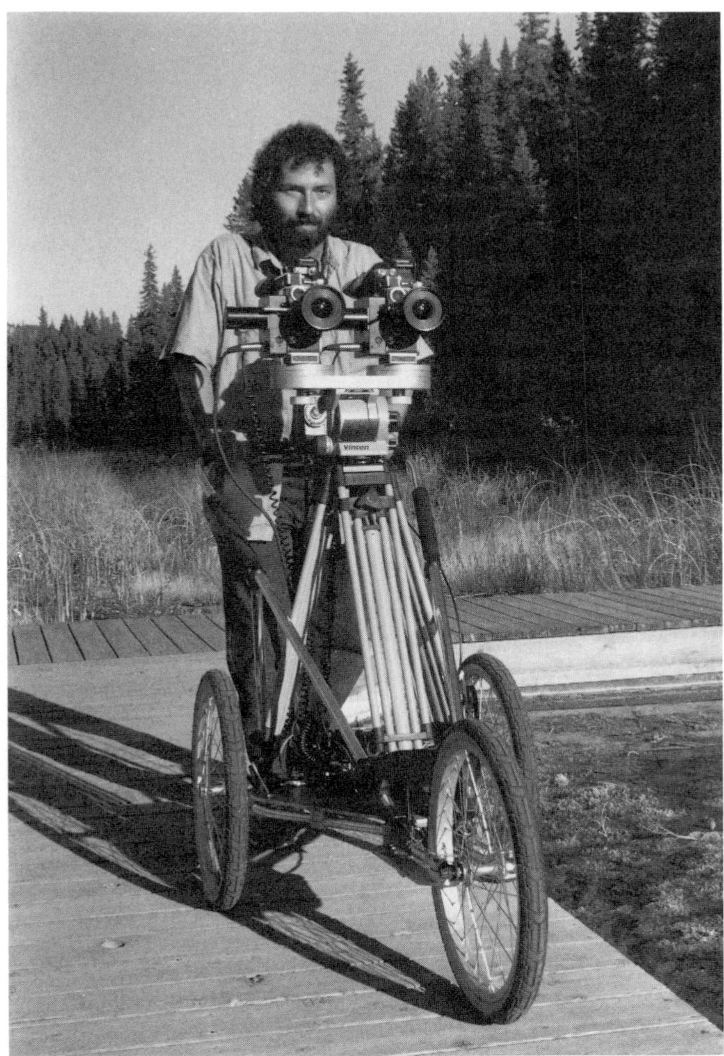

32. Michael Naimark with his camera rig for the stereogram images in *See Banff!* (1994). Photo Cheryl Bellows.

The cumulative effect of innumerable shades of tourists haunting the trails and icefields in virtual space and time is discomfiting; subjectively experienced unreal time reveals a supraindividual process that threatens the picture postcard perfection it has come to see. (Consider how firsthand experience of the fragile cave paintings of Lascaux has been replaced for visitors by a simulation.) Surrogate travel through a virtual environment may accelerate this dynamic of deterioration or, as some people fantasize, replace tourism in physical space entirely. Three of Naimark's time-lapse or what he calls "temporally undersampled"[24] segments are especially ghostlike and reminiscent of death: the spectral jogger on the Fenland Trail, the fleeting appearances of tourists on the Athabasca Glacier, and the thin stalks of grasses glowing translucent in the sun at Head-Smashed-In Buffalo Jump. These glimpses of images are perhaps as much a result of the instability of stereoscopic vision itself as of the haunting images in unreal time. However, it is more likely that cultural symbols (be they of self or other) will continue to be expressed across a continuum of cyberized "reality," theme parks, two-dimensional images, and virtual landscapes.

Naimark's strategy of showing the space around and in between the touristic vista can offer at times a sideways glance at an *image* of the natural world (not the material world itself) on its own terms (if such a view is possible), if only because the view is not the one expected. The artist shows us the glint on the lens and the halo effect of backlighting at Head-Smashed-In Buffalo Jump rather than the famous cliff itself; he shows us the rack-focus of the cyclone fence and spillway of Oldman Dam, not the spectacular river that created all the controversy; the abrupt end of a *z*-axis moves past picnic tables at the dam face, not the enormous cascade of water on the other side. "Expose the device!" seems to be a guiding principle of this piece, even if what we see exposed is a simulation. Most of all, the series reflects on the tourist's look itself as a trajectory and a subject position that we all occupy at some time or another, by repressing the touristic sight as much as possible from view. We are invited not to look at the "real" landscape, with all the cultural baggage it implies, but to see seeing as a historical and social practice figured in a virtual landscape. Naimark does not offer a project for controlling technologies of vision and thus of the natural world. Instead, this project marks how coarsely the material world is sifted through "realspace imaging" technology, how ill-fitting the landscape seems when pieced together by different historical imaging apparatuses, and how the underworld leaks through.

While the interface between the material and the virtual is the primary axis of Naimark's pieces, in Dove and Mackenzie's the underworld of dreams itself takes shape as a virtual world. Yet these different emphases hardly preclude the consideration of all other axes in any of the pieces; these axes are not meant to be exclusive, nor are they meant to be tools. They are ways of addressing different aspects of cyberspace. For instance, the virtual environment of *Archeology of a Mother Tongue,* like that of *See Banff!,* is also thematically concerned with the imbrication of the virtual and the material in a city that is like a gigantic prosthesis of the bodies of human beings who shape and are shaped by it. This thematic interpenetration of the virtual and the material is an exploration of the body in the city and the city in the body. In *Archeology of a Mother Tongue,* the body in question is that of a dead girl. As visitors, we will discover ourselves wandering within that body, in a post-mortem investigation of the traces of the city that are also allied with the cause of her death.

THE BODY, THE CITY—STEP INSIDE

The body of a dead girl is the point at which several narratives intersect; this body is also the pretext or enigma that drives the journey through four successive "envelopes" or worlds in Toni Dove and Michael Mackenzie's screen-based interactive artwork *Archeology of a Mother Tongue.* The opening premise of the piece is this: a coroner is flying to a foreign city in order to examine the child's body. Yet this is not a murder mystery nor is the enigma it poses to be answered by this one body (there are other bodies involved). This enigma is rather a lure toward a posthuman oedipal journey in which the question ultimately to be answered is not just where do I, or more accurately where does my body, come from, but also, am I human or machine, am I alive or dead?

Two different narratives are told by the voices of two persons or entities, a coroner (Toni Dove's voice) and a pathologist (Michael Mackenzie's); these narratives are actually about their own bodies as extensions of a virtual and

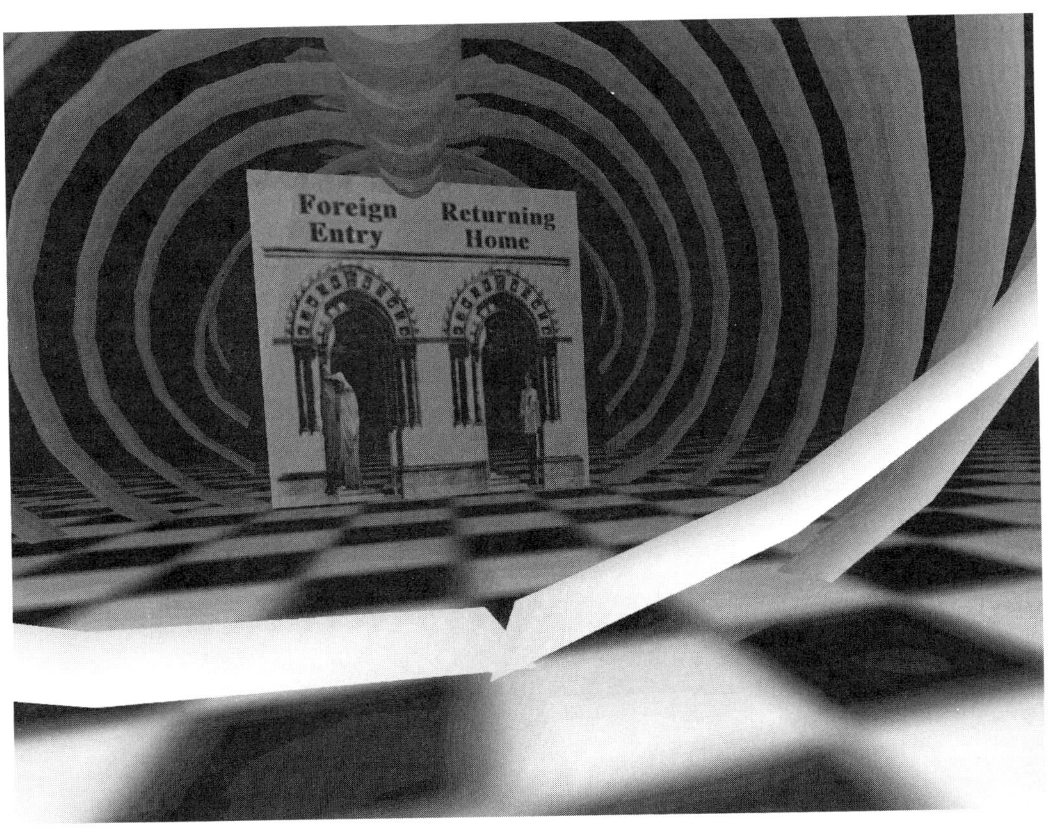

33. Toni Dove and Michael Mackenzie, *Archeology of a Mother Tongue* (1993), frame grab from one of three narrative "envelopes" or virtual environments. Digital image courtesy the artists.

material city. While this piece was a collaboration in every phase, Toni Dove was primarily responsible for the visual realizations and Michael Mackenzie for its plot(s) and interactive design. In addition, a team of programmers and computer-based designers, along with a powerful computer, contributed to this very labor-intensive project. As Mackenzie pointed out, while dramatic productions have a well-known division of labor and a general sense of the time necessary to complete each task, this is a new area in which the very production structures and practices must themselves be improvised. The degree of programming time this enterprise would involve was unanticipated; hence the completion date of the project did not coincide with a sense of having arrived at a completion of the piece or a solution to the problems it raised along the way. Toni Dove described the end result as an armature without the beta testing one would do from inside, much as Mackenzie sees it as a kind of drama that has been constructed and presented without the fine-tuning of a rehearsal process. Yet, far from being an orphan, this piece is perceived by both artists as an important part of the trajectory of their works, growing out of past concerns and enriching future work with a sense of what storytelling in a virtual environment involves.

Two of Toni Dove's earlier installations, *Mesmer: Secrets of the Human Frame* (1990) (which is, among other things, a retelling of the Dora story from Freud's famous case study, in her own voice) and *The Blessed Abyss: A Tale of Unmanageable Ecstasies* (1992), have many of the qualities of *Archeology,* including its glowing animated figures and subdued, ethereal color schemes. Even the wireframes of the computer graphic virtual environment have an otherworldly quality shared in her installations with visual elements from silent film frames and old anatomical drawings. The installations are screen-based and supported by real objects in what amounts to a kind of proscenium; however, the screen events are built from several narrative threads into an aural and visual envelope that is also a meditation on a range of cultural imagery and its assumptions. The sound tracks in all of these works share strands of multiple voices, male and female, interwoven and in counterpoint, spoke in different modes and persons and supported by layers of music and sounds that evoke the breathing and noises of the body. As in earlier work, the interior of the dead girl in *Archeology* is a secret and repressed space; her story is to be discovered through the pathologist's autopsy in the lab and the investigation of the coroner in the field, but most of all through the immersion of the visitor in a journey through the interior city of her virtual world.

The visual, aural, and thematic continuities between Toni Dove's installations and this virtual environment raise the issue of the difference between these artistic media in a larger sense.[25] While Dove's work underlines what these forms can share, the differences are also notable: a visitor who experiences both the installations and the virtual environment (especially if he or she is the person with the tracker that controls the point of view and direction of navigation in the virtual world) would presumably find them quite different. For one thing, while an installation is a composition of multiple, separate elements that may include images and objects, in a virtual world these are melded to become a seamless and dreamlike condensation in a purely virtual realm. The metaphysical consequences of immersion in such a seamless world can only be surmised at this point.

Another difference involves the capacity for interaction in the virtual environment. This is precisely what drew Michael Mackenzie to undertake a project in this new medium. A playwright and theater director for five years before embarking on this work, Mackenzie was intrigued by a possibility he considers as revolutionary as the one that originated Greek drama, when an actor speaking as a character disengaged himself from the chorus. Virtual environments take the extraordinary step of placing the audience in the position of the protagonist, granting it the power to make decisions that drive the plot to its once seemingly inevitable conclusion, which was once the preserve of the character in the play. Of course, in drama, these decision-making powers are actually the author's and they sometimes amount to character abuse. Mackenzie's 1989–1990 drama *Geometry in Venice,* staged in Toronto, confronts such an author, Henry James, as a distant and condescending character, with his own literate but impoverished creations in the setting of a premodern Venice as well as an impressionist Paris. A running theme in Mackenzie's plays is the cultural transition points associated with change in the dominant mode of representation: *The Baroness and the Pig,* performed in Prague in 1993, deals with a cultural transition associated with photography, while a play produced at the Banff Centre in 1994, *The Last Comedy,* reworks a transition between old and new comedy at a significant juncture in the history of the Golden Age of Athens. In *Archeology of a Mother Tongue* Mackenzie could experiment with the dramatic decision-making powers of the audience in a new medium and could encounter the consequences for the very notion of linear narrative and unfolding plot lines. It was also important to him to return to the theme of the city as a historical and social construct that marks itself into body and memory. In *Archeology,* however, that city is not a specific material city but a virtual condensation of many possible cities.

The multiple narrative strands of *Archeology* do not necessarily connect, though they are thematically related through the dead girl, and aligned through the ultimately linear trajectory of the piece in cyberspace. The visitor to the virtual landscape is also involved as a kind of story activator on his or her journey through the piece. The visitor chooses virtual objects (for example, a girl or a violin) by "touching" them using a dataglove, thereby determining the order of the release of narrative elements that float aurally above each envelope.

The narrative strands also take on different aspects of personhood: the voice of the coroner speaks subjectively as an "I," while the pathologist has two voices, one sounding like the graphic computer readouts that are visible on screen, and the other progressively more personal. For the visitor, the piece does not resolve whether it is offering the experience of a narrative or of the *act of narration.* In this screen-based piece a "driver" directs the point of view using a hand-held tracking device housed in a plastic toy camera, meanwhile interacting with screen objects by "touching" them by means of a data glove on the other hand. The audience is passive, along for the ride, not sure how far the point of view of the "driver" reflects the coroner's or another character's vision. Some visual segments are stored on laser disk, including the opening and closing image segments, plus "The Terminus." These offer a more cinematic than interactive experience, but they contain the spatial fluidity lacking in cut-and-piece cinematic space.

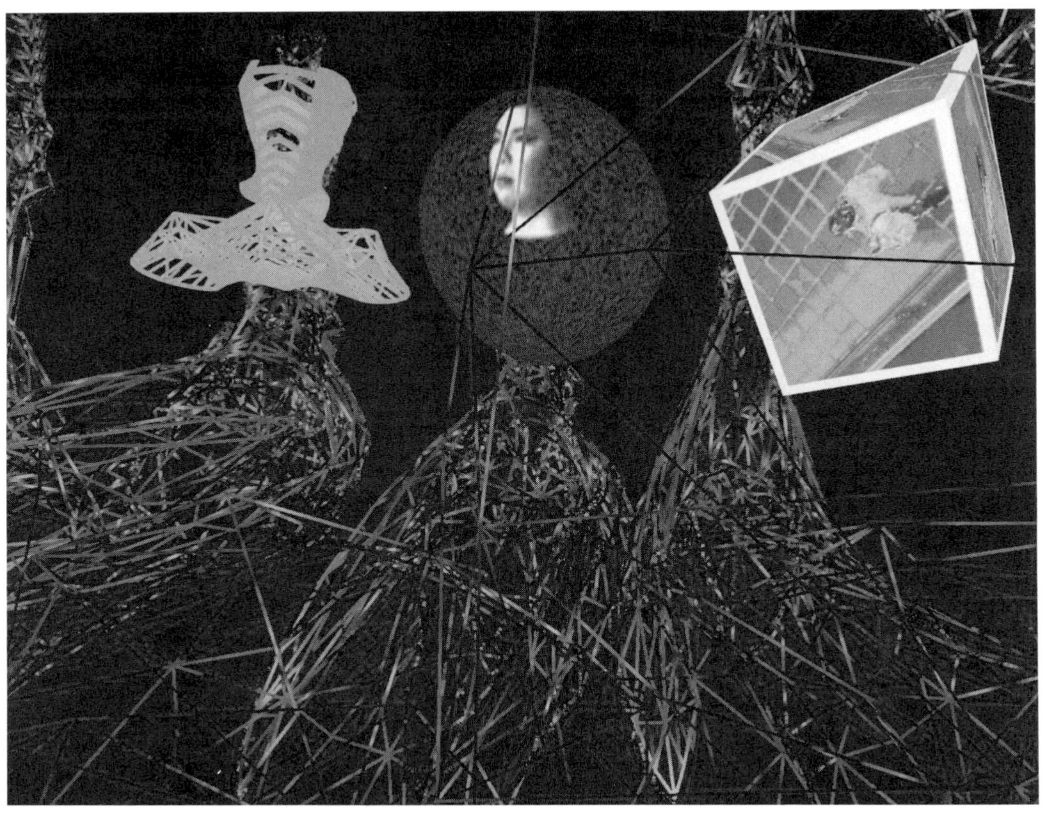

34. Toni Dove and Michael Mackenzie, *Archeology of a Mother Tongue* (1993), frame grab from one of three narrative "envelopes" or virtual environments. Digital image courtesy the artists.

Meanwhile, the images of girls or perhaps multiple Maris (the dead girl), their underwear worn on the outside like armor, pulsate angelically as spectral objects in what must be a parallel universe. One girl in a negative image plays a violin; another is a rotating head within a sales catalogue—she is for sale. It is in the verbal narration that we learn that Mari was a cybergang member, a nomad, a commodity on the adoption market, and, by the end of the piece, a child who died of trauma, but her body is decaying too fast to discover more. Throughout the piece there is a contradiction between the purity of the image world and the organic disintegration expressed in the aural world, a discrepancy that is explicitly addressed at one point in the narration.

Though the pathologist also speaks, the coroner's verbal story, initiated by her dream, dominates the first three envelopes. In the first envelope, "The Coroner Descends," the coroner's inner thoughts and memories are heard as she prepares for landing in a bleak foreign city and even, judging from the soundtrack, multiply foreign, for it includes an Arabic female voice and many other evocations of world beat aural elements). We hear that the city is full of cholera and other dangerous microbes, while we see, in the coroner's dream, that the city is composed of bright wireframes on the line of a Piranesi prison floating in the blackness of cyberspace.

Unlike a film narrative and like the material world, the wireframe landscape is continuously scanned by the "driver" without cuts or edits of space and time. Duration, or the pace of the narration of the story, is then very hard to control, depending as it does on the curiosity of the driver and his or her skill in releasing narration from objects in each envelope (all narration was stored digitally and triggered by programmed cues). On the other hand, the time it takes to pass from one envelope or world into the next is an instant, in a transition that nonetheless needs to be symbolically marked. The first passage is a drain hole or vortex that draws us into the second envelope, "The Coroner Arrives," in which appears the wireframe fuselage of an airplane that is also an enormous rib cage.

The passage to envelope three, "The Terminus," is a gate, which like the "silent house of sleep" in the *Aeneid* has two entrances: here the portals (to the

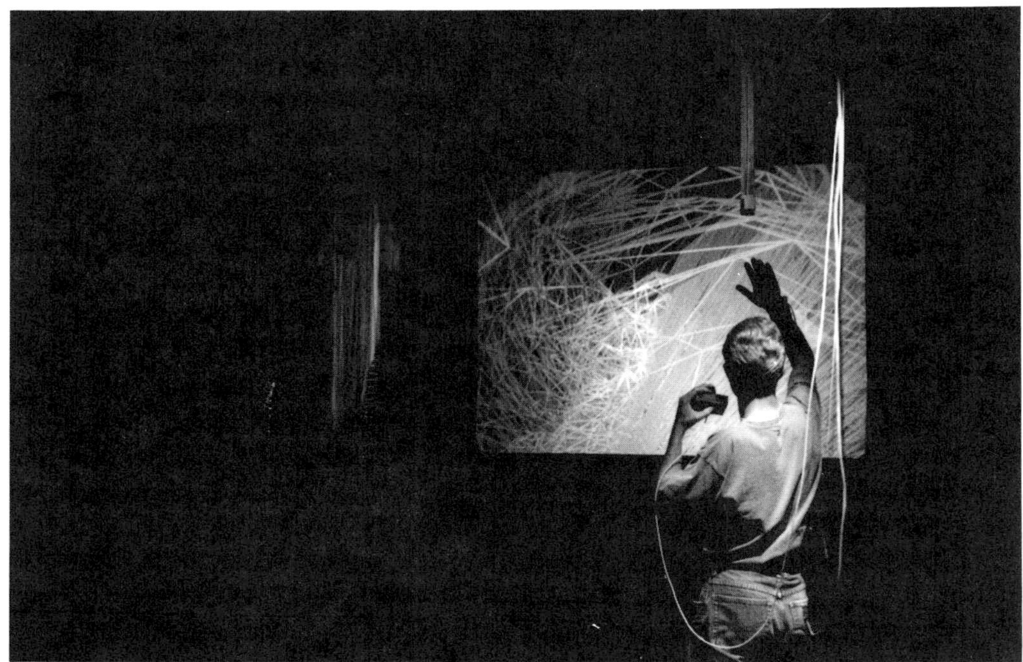

35. Toni Dove and Michael Mackenzie, *Archeology of a Mother Tongue* (1993), installation view with programmer Glen Fraser "driving" the piece. Photo Donald Lee.

city? to cyberspace?) divide natives and strangers. It is at this gate that the coroner is seen for the first and only time in the visual image, from overhead, passing successively in contrary directions. Which gate is hers? Her passage to and fro suggests an indecision about whether she is a stranger or a native, in what is almost an enactment of the *unheimlich*.[26]

The passage into the last envelope is a blackout or simulated power failure of the machine, which seems not only to maintain the city but to be the city. To push restart is to enter a different world. (As with a computer, there is short-term memory loss.) That fourth envelope, "The Pathologist," reveals an enormous wireframe skull–death's head. With that, the entire journey from envelope to envelope begins to resemble a journey through a wireframe city that is also an enormous corpse.[27] The fundamental situation is not unlike *Neuromancer,* in an even darker, bleaker mood. The body we travel through could be Mantegna's *Dead Christ.* Yet it is hard to ascribe the identity of the huge skeletal form to any one thing or being—it is a condensation of a pathologist, a city, a machine, the dead girl, and an entity with agency that is the virtual landscape itself.

The coroner's narrative concludes with a cognitive shift from alienation to a recognition of this enveloping machine-body-city as a kind of corporeal womb that produced her, only to put her up as a body for sale. The foreigner is actually coming home to a dark and deadly if no longer alien place, ripe for rebirth. The final narration (on laser disk, as was the opening segment) also tells us that the pathologist—and thus perhaps the city-body-cyberspace itself—is not a machine but human after all. (We are told the pathologist was promptly fired—unacceptable error level.) This narrational journey concerns the transformation of virtuality itself so that it might be something we recognize from our prehistory and someplace we might be at home. However, in the visual field, the skeleton of the virtual city remains fleshless and unredeemed, a dreamlike condensation in an aural envelope of voices.

At present, virtual environments are like packages that are designed without knowing what they might hold on the inside. The experience that can teach us how virtual environments are actually used is essential to understanding and situating such aesthetic experimentation. In a field where the lore of veterans is nonexistent and where conventions are invented ad hoc as one goes along, even the artists, as Toni Dove explained, could not be sure what to expect upon completion of the "machine" or environment they had spent months at the computer building, when they could step inside. Only after the piece was inaccessible to further tinkering and on display for a public could they assess how much of a difference it makes whether a visitor occupies the role of activator-driver or passenger, how much exposition was needed to make the piece gel, whether the variable order of segments within narrative envelopes changes the story itself in any significant way, what the appropriate pace is for the stories to unfold, and how interactivity affects timing. The massive computer power needed to drive the piece meant that it was erected for two brief periods at Banff as full-fledged cyberspace. Like other virtual worlds, this piece exists largely as a memory, as a video documentation, and as information on disk.

Conclusion: Landscape as Agency

The artists who produced the pieces discussed in this essay come from quite different cultural spheres: Michael Naimark's trajectory from the MIT Media Lab has been an oscillation between the worlds of art and media research. Toni Dove has worked through painting, music, and video installation before working with computer technology, while Michael Mackenzie was "in flight" as he put it, from a technological background in biotechnology and the economics surrounding the computer before he entered the theater world. Their projects at the Banff Centre are also primarily oriented around different axes of cyberspace, namely, the intersection of the material and virtual in Naimark's case and the entirely virtual world in Dove and Mackenzie's. Furthermore, each reflects on a different pole of cyberspace, from its use as a modeling and control mechanism for reality to its subjunctive uses in fantasy that subverts science by showing its links with the occult. Most certainly these pieces share a common concern with the body and with cyberspace in relation to death; yet, on further examination, the reader is compelled to ask: which cyberspace, which body, and which death?

As different as the sectors of cyberspace occupied by each piece were, the overarching themes of embodiment, agency, and death propose a common "structure of feeling," as Raymond Williams would term it, that seems to be based on features connected with the technology and the uncanny state of virtuality itself. There is a pervasive sense that a paradigm shift is under way and that subjectivity itself is in the process of transformation. Cyberspace is part of the shift in subjectivity as it presents the uncanny instance of agency that belongs to space. The virtual landscape is not just the ground or background or the landscape at which we look, but like Lacan's sardine can bobbing in the harbor it cannily looks at us. It tracks our every move and constitutes itself as a display in response to the indices of intention, the vectors of body position, gaze, and motion—that is, virtual space itself is *interactive*. The environment appears to be something "live" or animate, an agency "that we cannot accept as subject or persona in the traditional European sense, and which nonetheless constantly demonstrates that it sees us without revealing itself.[28]

Furthermore, the very notion of an environment or landscape, either virtual or real, becomes more and more difficult to distinguish from the symbolic field itself. We are immersed in symbols that have been lent not only agency but an agenda based on assumptions and viewpoints that are as relatively autonomous as the technology allows. The virtual environment fosters the illusion that we are in control, as if we were enunciating the very world before our eyes, in what amounts to performative speech acts. Or we may be led to rehearse a situation in which we are inside a technology over which we seem to have lost control, as in the theme park simulation or motion-control ride, only, of course, to come safely to rest four minutes later, our faith in machines renewed. However, the situation is actually far more complicated and out of our control as subjects. Where is authorship in the virtual environment? in the apparatus? the program? the performer? the animate environment itself? The nature of the virtual environment as a symbolic field or externalized imagination suggests why action within it is not free nor lacking in emotional and social consequences. Nor is one so completely disembodied in a fully immersive virtual world as is commonly believed: at minimum one retains a felt body that is mapped quite differently than the seen body—as Catherine Richards's *Spectral Bodies* video demonstrates. Furthermore, a puppetlike avatar is hooked up (but not in a one-to-one relation of effort to effect) to that spectral body in the virtual realm, awash in what amounts to a meaning system. (That is why a virtual persona can be raped[29] and why there is a relation between cyberdeath and psychic annihilation.[30]) Thus, the liminal and virtual realm in the machine is far from immune to moral issues or the ultimate questions of life and death.

Finally, although a virtual environment is an invention and a simulation, we cannot fully anticipate what it means to experience that realm until we are inside.

NOTES

1. Christopher Pinney, "Future Travel," in *Visualizing Theory: Selected Essays from Visual Anthropology Review 1990–94,* ed. Lucien Taylor (New York and London: Routledge, 1994), 423.

2. Ibid., 422.

3. Nor is the transfer from one level of cyberspace to another unproblematic. William Gibson's famous description of cyberspace—"Lines of lights ranged in the nonspace of the mind, clusters and constellations of data"—fudges the difference in level between data, digital objects, and their display, a difference that is more telling than one might anticipate if one wants to fill a virtual environment. (Of course, Gibson was using a typewriter at the time he wrote *Neuromancer,* and his fantasy of the space on the other side of the monitor was based on little knowledge of the state of computing. See Scott Rosenberg, "Cybervisionary," *San Francisco Examiner,* (August 7, 1994, C15). Timothy Binkley carefully distinguishes these levels in "Refiguring Culture," *Future Visions: New Technologies of the Screen,* ed. Philip Hayward and Tana Wollen (London: BFI, 1993), 92–122.

4. Published in New York by Schocken in 1975.

5. Pinney, "Future Travel."

6. Rosenberg, "Cybervisionary."

7. N. Katherine Hayles's distinction between absence/presence and randomness/pattern, made in "The Seductions of Cyberspace" in *Rethinking Technologies,* ed. Verena Andermatt Conley (Minneapolis: University of Minnesota Press, 1993), 173–190, especially 186f., might be conceived as designating different stages of information (which have different psychic roots and subject-effects) that coexist rather than that succeed each other in a developed information society.

8. Jeffrey Schulz, "Virtu-Real Space: Information Technologies and the Politics of Consciousness," *Computer Graphics: Visual Proceedings* (ACM Siggraph, Annual Conference Series, 1993), 159–163.

9. I discuss the built environment and commodities as part of a system of the exchange of images that pass through different reality statuses in "An Ontology of Everyday Distraction: The Freeway, the Mall and Television," in *Logics of Television: Essays in Cultural Criticism,* ed. Patricia Mellencamp (Bloomington and London: Indiana University Press and BFI, 1990), 193–221.

10. In "Art in Cyberspace" (published as "¿Ciberia o comunidad virtual? Arte y ciberespacio," *Revista de Occidente* 193 [February 1994], 73–90), I developed an earlier version of these three axes.

11. A point made by Leroy Littlebear during the Fourth International Conference on Cyberspace at the Banff Centre for the Arts, Banff, Alberta. Wolfgang Schivelbusch deals with the history of street lighting and its social and cultural ramifications in *Disenchanted Night: The Industrialization of Light in the Nineteenth Century,* trans. Angela Davies (Berkeley: University of California Press, 1988).

12. Hayles, "The Seductions of Cyberspace," 173. The issue of computers and control was discussed in Bill Nichols, "The Work of Culture in the Age of Cybernetic Systems," *Screen* 29 (Winter 1988), 22–46.

13. Recent criticism of interactive art by the author appears in "Art in Cyberspace:

Interacting with Machines as Art at Siggraph's 'Machine Culture—The Virtual Frontier,'" *Video Networks* 17:5 (October/November 1993), 19–23. An appreciation of the acculturation of the machine to the human appears in "Judith Barry: The Body in Space," *Art in America* (April 1993), 116–121, 143, while a critique of the accommodation of the human to the machine in terms of the cyborg can be found in "What Do Cyborgs Eat? Oral Logic in an Information Society," *Discourse* 16:3 (Spring 1994), 86–123, and in *Culture on the Brink,* a collection of essays published by Bay Press and the DIA Foundation in 1994. Another discussion of the mixture of the physical and the projected appears in "Muntadas' Media-Architectural Installations," *Muntadas: Trabajos Recientes* (Valencia, Spain: IVAM Centre Julio Gonzalez, 1992), 112–117.

14. I discuss the enigma of virtual space and networks as "spaces" at greater length in "Enthralling Spaces: The Aesthetics of Virtual Environments" in *ISEA 94: The Fifth International Symposium on Electronic Art* (Helsinki: University of Art and Design, 1994), 83–89.

15. Brian McHale, *Postmodernist Fiction* (New York: Methuen, 1987).

16. William Gibson, *Neuromancer* (New York: Berkeley, 1984), 244.

17. Marsha Kinder of the Department of Critical Studies in Cinema-Television at the University of Southern California developed the discomfiting gender implications of the killing moves and the arcade milieu of *Mortal Kombat* in a paper delivered at the Console-ing Passions Conference sponsored by the University of Arizona in Tucson in April 1994.

18. Allucquère Roseanne Stone discusses the notion of warrant and the "fiduciary subject" in "Sex, Death and Architecture," *any 3: Electrotecture: Architecture and Electronic Culture* (November/December 1993), 34–39, among other places.

19. Ivan Sutherland, "The Ultimate Display," Information Processing Techniques Office, ARPA, OSD, published in *Information Processing, 1965: Proceedings of IFIP Congress* 65 (Washington, D.C.: Spartan Books, 1965), 508.

20. The passage continues: "There were no signs directing your gaze, no coin-operated binoculars, and no brochures answering your unasked questions about local flora, geology or the history of the land. . . . By the mid-twentieth century, it seemed, nature had to be explained to its human inhabitants; it was not enough to just try to experience it. As a result, conflicting information about the natural world blankets out visual and aural environments. Much of this information is promotional—that is to say, often misleading, mystifying, or simply irrelevant." Alexander Wilson, *The Culture of Nature: North American Landscape from Disney to the Exxon Valdez* (Oxford, England, and Cambridge, Mass.: Blackwell, 1992), 53.

21. Tourism and its practices have been the object of cautionary and demystifying sociological and ethnographic studies and semiotic analyses for at least two decades. (It is amazing how few positive values are accorded to the romance and the revenue involved.) For instance, the "self-destruct theory" of mass tourism puts ostensibly harmless activities such as sightseeing in the context of a slow death of a cultural landscape, while "Doxey's index of irritation" charts the phases of the host population's reaction to guests, from initial euphoria to antagonism. Gareth Shaw and Allan M. Williams's, *Critical Issues in Tourism: A Geographical Perspective* (Oxford, England, and Cambridge, Mass.: Blackwell, 1993) reviews many of these theories and models. Thanks to Steve Seid for drawing my attention to this reference.

22. Dean MacCannell's *The Tourist: A New Theory of the Leisure Class* (New York: Schocken, 1976) proposes that the loss of authenticity in everyday life fuels the tourist's search for authentic nature and community, thereby producing the simulation of both and the recedence of the physical world under superimposed signs and images.

23. Homi Bhaba, "Interrogating Identity," in *Identity,* ed. Homi Bhaba (London: Institute of Contemporary Arts, 1987), 6, cited in Pinney, "Future Travel," 422. Pinney questions whether surrogate travel is travel at all if it does not leave the spot. However, this understanding of the virtual environment does not take into account that the display and interactive devices do indeed act as frame and that virtual motion is travel in effect, if not actually. The distinction between actual and virtual travel becomes less telling once we recognize that "depth" in the narrational journey is a metaphor. Therefore, why can movement through the space of an embodied metaphor not be as effective as movement through physical space?

24. Michael Naimark, "Frame Rate," *Elements of Realspace Imaging,* Technical Report of the Apple Multimedia Lab, San Francisco, 1991, 34f.

25. I discuss the aesthetics of video installation in "Video Installation Art: The Body, the Image and the Space-in-Between," in *Illuminating Video: An Essential Guide to Video Art,* ed. Doug Hall and Sally Jo Fifer (New York: Aperture and BAVC, 1990), 153–167.

26. Sigmund Freud described the "uncanny" as a both strange and familiar feeling that recalls the moment of the discovery of sexual difference or the sight of a corpse. It is a feeling of the unfamiliar or "not" home that is nonetheless "home," familiar from early childhood experiences.

27. Elizabeth Grosz, "Bodies-Cities," in *Sexuality and Space* (Princeton: Princeton University School of Architecture, 1992), 241–253, deals with the city *and* the body (not city as body). Susan Stewart's *On Longing: Narratives of the Miniature, the Gigantic, the Souvenir, the Collection* (Baltimore: Johns Hopkins University Press, 1984) is useful on the imaginary social body as one gigantic individual. In a particular example of the pathetic fallacy, the landscape in Ernest Hemingway's "Hills Like White Elephants" evokes the pregnant body of the protagonist. The city as skeleton or corpse evokes the symbolic or "unfilled" nature of cyberspace, though it may also be a bleak view of contemporary cities.

28. Friedrich Kittler, "A Conversation between Peter Weibel and Friedrich Kittler," in *On Justifying the Hypothetical Nature of Art and the Non-Identicality with the Object World* (Cologne: Robert Fleck, 1992), 162.

29. Julian Dibbell, "A Rape in Cyberspace," *Village Voice* (21 December 1993), 36–42.

30. See Howard Rheingold, *The Virtual Community: Homesteading on the Electronic Frontier* (Reading, Mass.: Addison-Wesley, 1993), 32–37.

Time Traveling in the Gallery: An Archeological Approach in Media Art

Erkki Huhtamo

Zoëtropes, kinetoscopes, stereoscopes, phonographs, pinball machines, fortune-telling machines, periscopes and guns, household appliances, vintage television sets, even obsolete computer displays and videogames—the whole repertoire of old tech, rapidly increasing due to continuous technological change, seems to have invaded the art gallery. Visiting a media art exhibition today may feel like visiting a museum of technology, at first sight. Of course there are, and should be, differences between a show of contemporary art and an educational display. The meticulously crafted interactive replicas of precinematic devices in places like the Museum of the Moving Image in London, or the inviting multitudes of knobs, handles, joysticks, and other interface devices made available for the public at science centers such as the Exploratorium in San Francisco and La Villette in Paris, may evoke the works and strategies of "archeologically" inclined media artists, both in form and in modus operandi. Significant technological artworks have, in fact, been commissioned by these institutions.[1] Without a doubt, designing an ingenious gadget that will offer the visitor the phenomenological experience of traveling back in time to another technological era and/or having direct (often tactile) access to an application kept usually "behind the scenes" is an important educational goal. It does not, however, automatically confer upon the piece the status of art. Something else is required; this something is now being provided by an increasing number of artists.

The lives of machines from the past are being relived by media artists dealing with digital and interactive technologies to a striking extent.[2] Although artists like Jeffrey Shaw and Toshio Iwai have worked in the "archeological idiom" since the 1980s (and even earlier), the emergence of what I will call the *archeological approach* in media art has primarily emerged, as a wider aesthetic concern, in the 1990s. Noted artists such as Rebecca Cummins, Paul DeMarinis, Ken Feingold, Lynn Hershman, Perry Hoberman, Heidi Kumao, Michael Naimark, Catherine Richards, Jill Scott, and VNS Matrix, among others, are producing artworks that incorporate explicit references to machines from earlier phases in the development of technoculture. These incorporations form major aesthetic and structural elements of their strategies.

In some cases, the appropriation of the formal attire of old tech reaches a point at which the artwork "asks" to be interpreted as a ready-made or a remake of an original (allowing us to classify for example Catherine Richards's installation *The Virtual Body* as a "peep-show machine" of a kind). More often, however, the explicit historical reference point is willfully displaced and mixed with other associations. The result is a hybrid construction, which transcends the historical referent (without effacing it) and presents itself as a deliberately ambiguous spatiotemporal and parahistorical object. Paul DeMarinis's sound installation *The Edison Effect* is a series of machine-objects that investigates different forms of sound reproduction, from Edison's phonograph cylinders to recent laser technology. But instead of organizing them chronologically, DeMarinis juxtaposes his references or piles them in a highly ambiguous manner. In one of the objects, titled *Fragments from Jericho* (1991), he uses a laser beam to produce sound from a clay pot rotating on a turntable, poetically and hypothetically extending the era of sound reproduction by thousands of years.

Other archeological art projects make no attempt to simulate a gadget from the past but adopt a more oblique approach. Ken Feingold's *Childhood/Hot and Cold Wars (The Appearance of Nature)* (1993), for example, is concerned with the influence of television in the 1950s on the American collective discursive system of hopes, fears, and desires. Feingold has combined disparate elements with their own associative potential (parts of a fifties suburban home, a replica of the "A-Bomb house" skeleton in Hiroshima, a grandfather clock, hours of 1950s television clips animated by turning a transparent globe, an aluminum framework) into an evocative "sculpture."[3] In yet other cases, the archeological

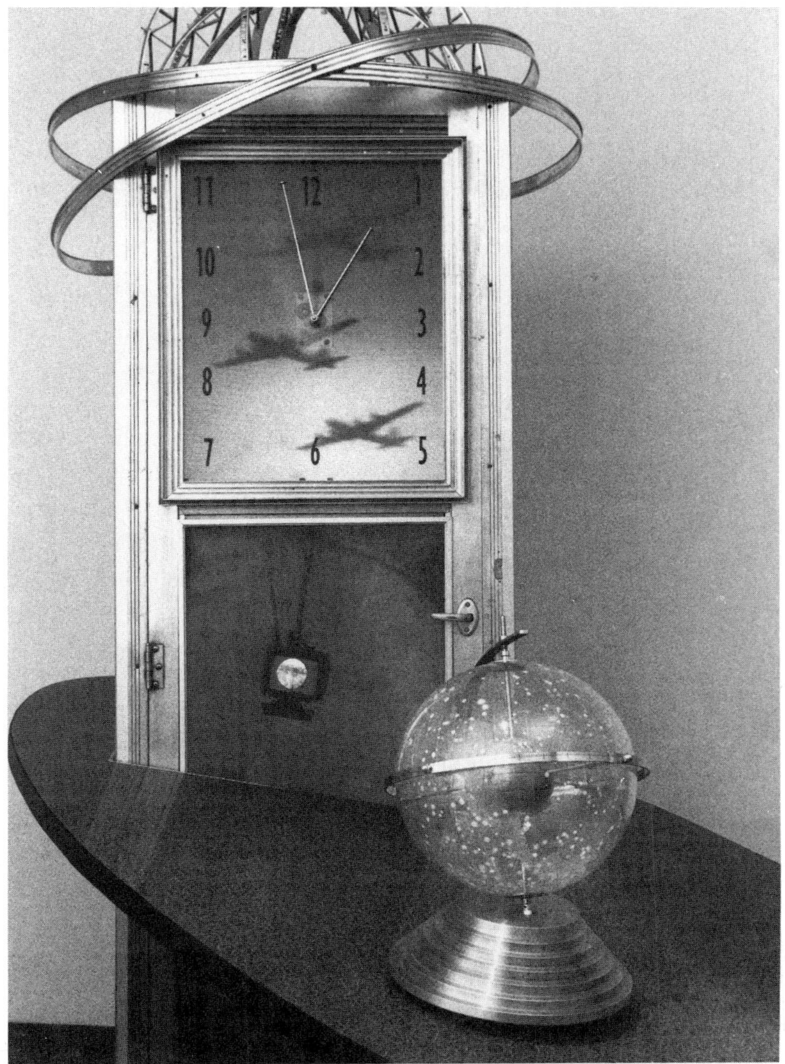

36. Ken Feingold, *Childhood/Hot & Cold Wars (The Appearance of Nature)* (1993), interactive sculpture.

concern has found an outlet that does not directly evoke the forms (and thus the phenomenological experience) of old tech. Contemporary technology has been used as both the terrain and the tool for media archeological excavation in works such as the CD-ROMs produced by Christine Tamblyn (*She Loves It, She Loves It Not: Women and Technology,* 1993) and by James Petrillo (*Cinema Volta,* 1993). Both present a mixture of personal reflections, historical material, and cultural and ideological discourses about technology, its development, and its position(s) for the user to excavate.[4]

The purpose of this article is to take a closer look at the phenomenon delineated above, which I have named the archeological approach in media art. Many questions can be raised about the structures and content shaping this approach: How does it relate to the general development of the field of art and technology? Is there something "new" in it, something not explored earlier? How can we explain the emergence of such an approach in the present cultural context? Even at the risk of falling into the "intentional fallacy" trap, one is also tempted to ponder the kinds of significations the artists themselves attach to their involvement with old tech. Are there similarities or differences between their approaches? Is the archeological tendency homogeneous or fragmented? In other words, my aim is to find a set of interpretative "frames" for the phenomenon in question, while trying to make sense of its "nature."

FROM THE ARTIST-ENGINEER TO THE ARTIST-ARCHEOLOGIST

Beginning with the classic avant-garde movements of the early twentieth century (or even earlier, depending on the perceived relationship among art, science, and technology)[5] the "machine in the gallery" has been a central concern for artists working with new technologies (and even for many of those using traditional tools). The creations of Marcel Duchamp, Man Ray, László Moholy-Nagy, Vladimir Tatlin, Alexander Calder, and many others laid the groundwork for the artistic investigation of issues related to technology, culture, and society. These issues can be presented as sets of binary oppositions: individual versus industrial creation, the artist versus the engineer, "art" versus "life," traditionalism versus progress, the blessings versus the destructiveness of technology, anthropocentrism versus the inhuman character of the machine, "art for art's sake" versus applied art and commercialism. One effect of the continuous artistic exploration of this field has been the gradual merging of some of these oppositions. This does not mean, however, that they have been annihilated altogether.

The issues raised by the classical avant-garde were picked up and expanded by the "neo-avant-garde" movements of the 1950s and 1960s. In spite of the profuse material produced by these movements—exhaustively explored by Frank Popper and Jack Burnham—the attitudes toward the machine seemed to fall within the paradigm established earlier, characterized by the division between dadaism and constructivism.[6] Among the neo-avant-garde the former tendency, best represented by the Fluxus movement, launched a critical and skeptical reaction against machine culture by appropriating its products, displacing them from their everyday contexts, and "preparing" and maltreating them. The ensuing Fluxus machines were anti-machines, deliberately "inpoverished" hybrids meant to shock and satirize the stale matter-of-factness of the industrial high-tech society. But they had a romantic air as well, showing a love for kitsch and a longing for a society of "humanized" technology where personal tinkering with machines would have replaced the anonymity of industrial production. This strangely echoed the preindustrial ideals about craftsmanship and cottage industry. The Fluxus machines thus reflected a general cultural uneasiness about technology—formulated in 1959 by C. P. Snow in his model about the incompatibility of the two cultures, the scientific and the non-scientific—rather than resolved it.[7]

The latter tendency, represented by kinetic art, early computer art, and forms of light art, for example, aimed at resolving the split between traditional artistic creativity and scientific and industrial forms of creation. By adopting state-of-the-art technologies and applying mathematical and scientific models (cybernetics, information theory, structuralism), artist-engineers attempted to reveal the creative potential lying hidden in these technologies. Such works could sometimes also be seen as demonstrations of certain scientific phenomena.

Ben F. Laposky, who started producing his "oscillons" or "electronic abstractions" on an analog computer in the early 1950s (thus pioneering the field of computer imaging), described his creations as "an excellent example of the possibility of employing modern technology in art and of demonstrating a relationship between science and art. They are also visual manifestations of some of the basic invisible aspects of nature, such as the movement of electrons and energy fields."[8]

The technological artist came to be considered as much a system builder as a creator of his own personal artwork. Some manifestations of this were the "responsive" cybernetic sculptures created by Nicholas Schöffer, Wen-Ying Tsai, James Seawright, and others, which provided early sources of interactive art. Even more important in creating the role of technological artist was the activity of Myron W. Krueger, who bridged the gap between computer science and artistic activity both in his personal development and in his art. In the late 1960s Krueger started developing responsive systems-cum-environments-cum-artworks (*Glowflow, Metaplay, Videoplace*), which directly engaged the spectator, bringing the question of the human-machine relationship to the forefront of technological art.[9]

These pioneers tackled the task of tinkering with new technologies to turn them into "tools" for producing experiences that differed from those offered by the established media and the military-industrial complex. This happened in a "constructive" spirit, even though the ultimate goal of such activities varied. Should the new forms and applications be introduced to the dominant society with the aim of changing it from the inside (in the form of industrial design, for example), as Moholy-Nagy suggested in the 1920s, or would it be better to keep them deliberately in the margin, contributing to the changing of the society only gradually by building alternative media realities and developing new aesthetic sensibilities? In any case, science and technology were considered an immense storehouse of machine parts, theories, and natural laws full of unused (or abused) creative potential. Technology could be turned into artworks, or into alternative realities, or into such wonderful "instruments" as the early video synthesizers built and designed by independent artist-experimenters Eric Siegel,

Stephen Beck, Bill Etra and Steve Rutt, and others.[10] Last but not least, of course, it could be turned into a vision of itself, projecting technology as the panacea of the future.

Although the exploratory and demonstrative aspect of technological art remains an important goal (and is still pursued by pioneers like Myron Krueger, Peter Weibel, and Woody and Steina Vasulka, as well as younger artists such as Karl Sims and Michael Saup), it has been balanced by the gradual coming of the archeological approach.[11] This means that the activity of creative tinkering with technology has been placed within a wider frame of reference that emphasizes the historical, social, and cultural determinants of technology instead of merely the scientific ones. Any form of technology is seen as a bearer of significations, which precede it, mold it, and guide its development. In other words, the artistic approach to technology has now less to do with concrete artifacts, and with the ideologically "pure," "scientific" theories informing them, than with the *discursive formations* enveloping them. This concerns not only the institutionally established and legitimized applications of technology, or *cultural forms* (as described by Raymond Williams), but unrealized projects (as expressed in patent applications, for example) and "discursive inventions" (machines that exist only as discourses, not in concrete material form) as well.[12]

Archeologically oriented artists do explore new technologies, but they simultaneously take a keen interest in using their art as terrain for confrontations with technology-related ideological issues involving class, gender, power, the historical nature of vision, and the relationship between high and low culture. Instead of forming a kind of double focus, these approaches coincide (or are at least meant to do so), as in the case of successful stereoscopic vision. This is proven by the ways artists have been dealing with one of the key issues of interactive technology, that of interfacing the human and the machine. Although the concept of "interface"—in the sense of a "connecting surface" between the human and the machine—originated with computer science and is mainly associated with human-computer interaction, archeologically oriented artists have shown that fundamental problems go further back in history and are deeply intertwined with ideological issues.[13]

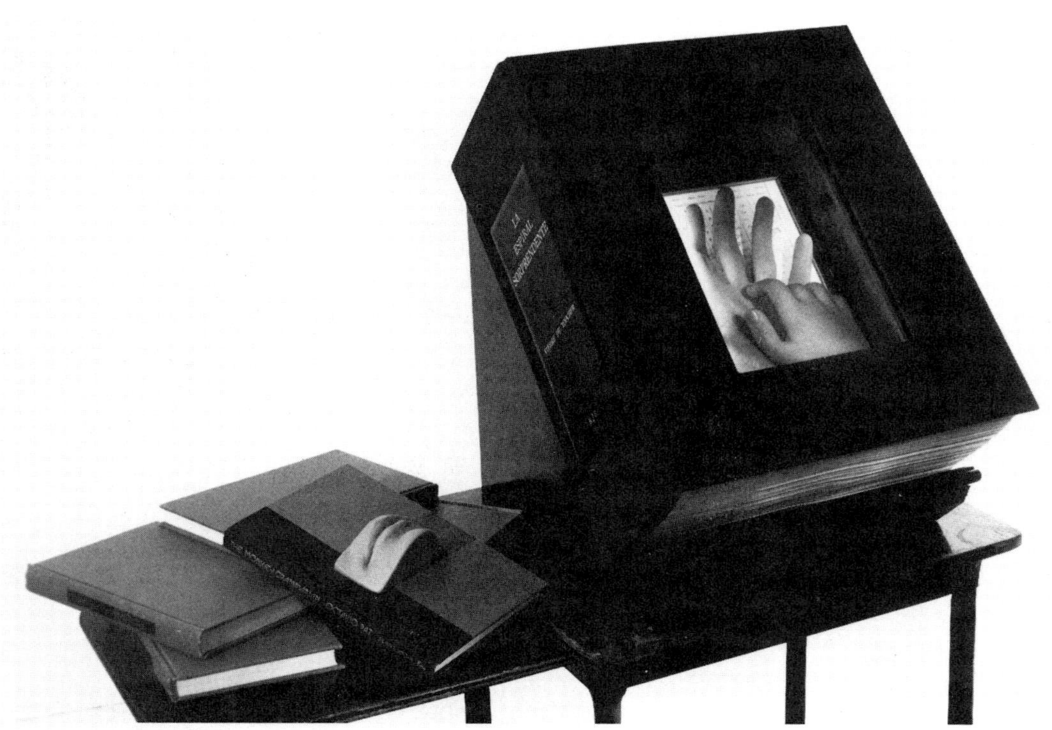

37. Ken Feingold, *The Surprising Spiral* (1991), detail of interface objects.

Both Lynn Hershman and Ken Feingold have explored the metaphor of "tactility," now a central aspect of the field of interface design and crystallized in the idea of the "touch-screen." Hershman confronted the sexual and gendered implications around the tabus of touching (and even "penetrating") an image in her interactive videodisk installation *Deep Contact* (1990). In his installation *Surprising Spiral* (1991), Ken Feingold approached the question in another way by reenacting the ways anthropomorphic interfaces (simulated hands or feet, for example) had been used, first, in old arcade machines to mediate the contact with technology and, then, by Marcel Duchamp and the surrealists to question the "untouchable" quality of classical art and the ideology it embodies. Both Hershman and Feingold developed ingenious interfaces as access points to their pieces—Hershman a touch-screen with a graphic representation of a woman's body, Feingold an "interactive book" with a transparent touch-screen cover and simulated hands visible beneath.[14]

Perry Hoberman's *Faraday's Garden* (versions since 1990), a kind of technological sculpture garden consisting of found household appliances, home movie projectors, and the like, shows yet another approach. Hoberman has made the gadgets, most of them electric "prostheses" or "extensions" of the human hand, work "independently" (actually triggered by sensors under the visitors' feet). The simple gesture of eliminating the hand-mediated contact is rich in implications. It makes one think about the modernist ideal of the "automated home" (prevalent in the 1950s and 1960s, the era of origin of most of the gadgets) and its relationship to the subsequent ideology of interactivity. *Faraday's Garden* also suggests the theme of anthropomorphism in machine culture by implying that the gadgets are somehow "alive"—at least semi-independent creatures one easily associates with both the Frankensteinian tradition and the recent discussions about artificial intelligence and interface agents.

Finally, Hoberman's work reminds us that we should not draw a priori distinctions between "media machines" and those meant for killing or cleaning or cooking. The functions of these machines may at first sight seem quite different, but from the point of view of interfacing humans and machines they may have surprising similarities. All these machines, mediated by the strategies

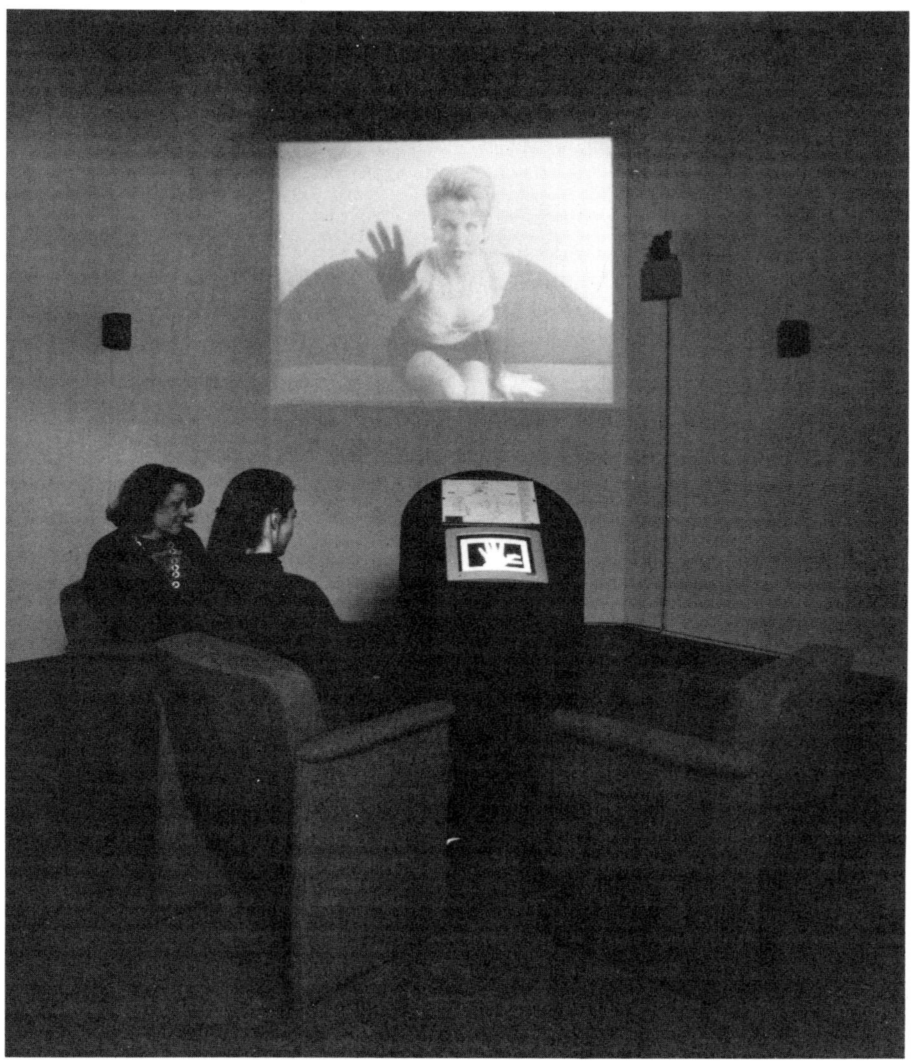

38. Lynn Hershman, *Deep Contact* (1990), installation view. Photo Ben Blackwell. Courtesy the artist and Robert Koch Gallery, San Francisco.

and discourses of consumer culture, constitute fairly tenacious subject positions that have strongly, and in often unnoticed ways, influenced how we see the world and order our everyday experience of space and time. Considering the history of women's uneasy relationship with technology in a male-dominated society, it is no wonder that the technological zone spanning the office, the telephone exchange, and the home has been excavated by artists with a feminist orientation, for example Christine Tamblyn, Heidi Kumao, and Jill Scott (for example in her ambitious installation *Machinedreams,* 1991–1994).[15]

To sum up, the role of the artist-engineer, which rose into prominence in the 1960s (although its two sides rarely met in one person), has at least partly been supplanted by that of the artist-archeologist. To a certain extent this can be explained by the merging of features from earlier trends in art and technology. The anarchistic critique of the machine as a cultural form launched by the dadaists and some Fluxus artists has been domesticated and married to a more constructive approach. This approach tries to maintain a critical attitude toward technoculture but contributes simultaneously to its development by inventing new forms of information storage and retrieval, as well as designing new modes of interaction. Few archeological artworks use shock treatment or (simulated) blows below the belt to drive their issue; most of them choose the way of creative dialogue with the user, which further emphasizes the role of successful interface design.[16]

Virtual Traveling in Time

With the coming of the archeological approach, the gaze of the media artist that was directed primarily toward the future has now been supplanted, or rather supported, by another one that faces the past. This gaze is not, however, motivated by the nostalgia of a techno-buff, or the postmodern transfiguration of the banal as in Jeff Koons's "prepared" found appliances. As I hope to make clear, the activity of this gaze can be seen as an attempt to go *beyond* postmodernism, to initiate a dialogue with the past to counter the constant blurring of boundaries and definitions that is characteristic of the "postmodern condition" and largely a product of the spreading of audiovisuality. Yet, although the gaze of the media archeologist faces the past, it is not fixed; rather, it is extremely mobile. It scans incessantly the historical panorama of technocultural forms, moving back and forth in time, looking for correspondences and points of rupture. In the end it (re)turns to the present and, eventually, the future.

There is, however yet another gaze that is a *constructed feature* of the archeological artwork itself. This gaze exists as a series of proposed subject positions for the user implanted into the structure of the work; while these may be "preferred," they by no means exhaust the ways in which the work can be read. One of the peculiarities of the media art installation is its Janus-faced character as both an object and a process (referring to the presence of the time-based element, video or computer images, and in the case of interactive artworks to the "negotiation" of meanings between the user and the work). The constructed gaze is split between these "dimensions," which I will call the "outer form" (the spatial, material aspect of the installation) and the "inner form" (the virtual worlds hidden "within" it and as if entered via the artwork). In an archeological artwork, switching between these dimensions, as if constantly changing focus, means alternating between different subject positions, and also "traveling" between layers of time.

Judging from its outer form, Michael Naimark's *See Banff!* (1994), for example, evokes something "antique" (it is actually a hybrid of a kinetoscope and a mutoscope, a hand-cranked peep show device for viewing moving images). By peering into an eyepiece and cranking a handle we see stereoscopic film sequences unfold, mostly landscapes and touristic sites (shot by Naimark using time-lapse cinematography in and around Banff National Park in Canada). Many of the forty sequences available are traveling shots taken along trails; others show landscapes from a moving vehicle. The deliberately jerky movements may evoke the experience of early motion pictures (indeed, most of the cinematic figures Naimark uses reenact the geographical "views" shot by the Edison company or the Lumière brothers), which is "confirmed" by the silent film–like captions. On the other hand, we are clearly watching contemporary scenes related to the rituals of tourism, which connotes amateur film and its aesthetics.

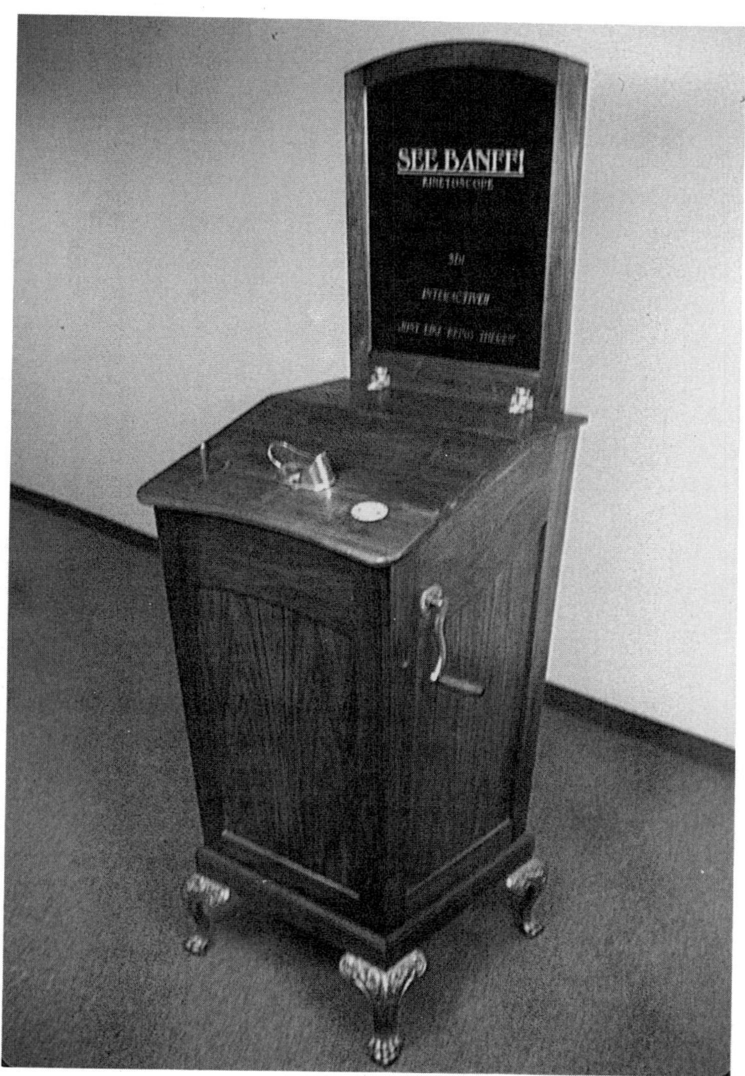

39. Michael Naimark, *See Banff!* (1994). Photo courtesy the artist.

Again, the changes in "image density," and consequently in the rhythm patterns of the sequences, highlight the formal experimentation underlying the enterprise. The result is a highly layered construction, in which the simulated gaze of the silent-film subject interacts with the (equally simulated) touristic gaze and with yet another one related to the formal experimentation. Those aware of Naimark's earlier work will recognize his long-time interest in representing real space and transferring it to another system of signification.[17]

Similarly, Jeffrey Shaw's installation *Inventer la terre* (1986), commissioned by the La Villette science center in Paris, is a densely layered cluster of history, myth, and power. The visitor first sees a chrome-plated column standing on a round black terrazzo pedestal, inlaid with brass signs: the cabalistic zodiac. Getting closer, one notices that the column is a kind of periscope, with an eyepiece and two handles, one on either side. By grabbing the handles, the visitor can turn the column around; simultaneously gazing into the eyepiece reveals a "see-through" panoramic view of the surrounding science center. Yet there is another panorama superimposed on this one—a string of computer-generated views of creation myths, concepts of the universe by different cultures. The images seem immaterial, like mirages floating in the air. By pressing little knobs in the handles, as if launching a torpedo, the visitor can animate the views, "visit" them.

The outer form of Shaw's installation may evoke an ancient cosmological monument—the stones of Carnac, or Stonehenge—a site of ideological domination, of an "absolute" truth about the universe. This interpretation is doubled by the fact that the work is also a panopticon, a "machine of vision" with unrestricted scopic power built into its construction. The user who immerses his or her eyes into the eyepiece gains a "panoptic gaze" for surveying the science center, with all its pointers to high tech. This evokes the close connection between electro-optical surveillance and the contemporary technoculture. Yet the superimposed images, floating in the field of vision, add another meaning, achieving a kind of *détournement.* Instead of launching a torpedo we activate different ideological discourses, none of which is presented as superior to the other. We thus launch ourselves on virtual voyages of discovery in mythical-

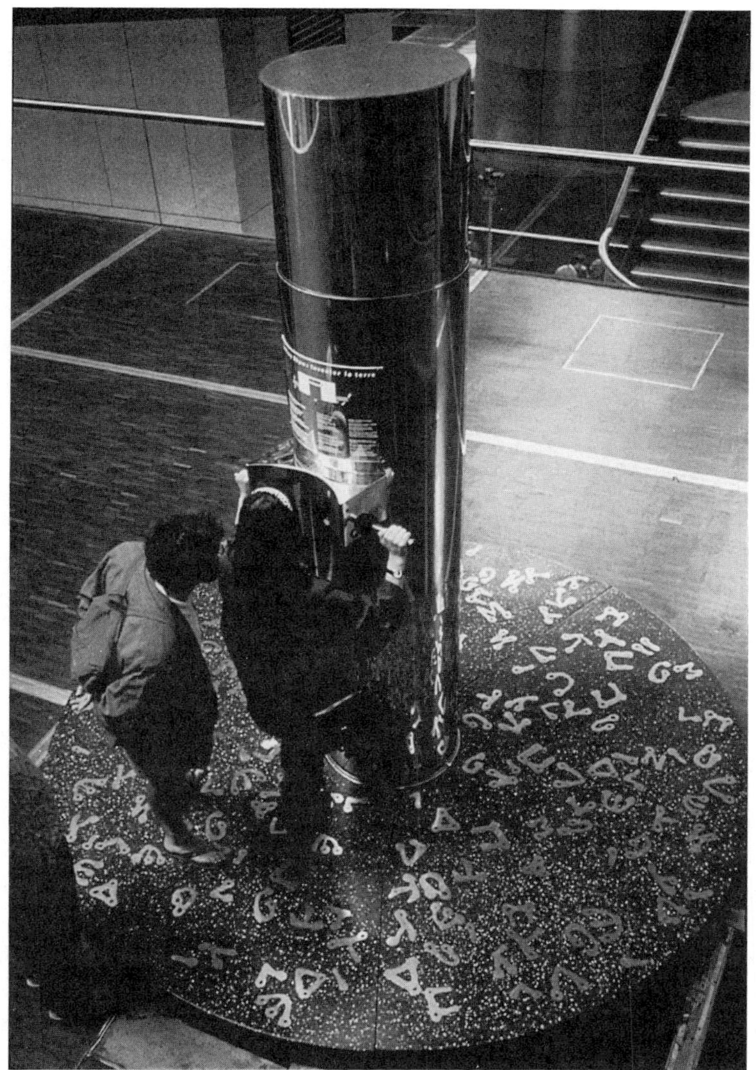

40. Jeffrey Shaw, *Inventer la terre* (1986).

conceptual time. These become "philosophical journeys" in the tradition of Montaigne or Goethe. *Inventer la terre* makes us encounter the plurality of the world, not to force it into a preexisting formula but to explore its variety, and to broaden our ideological horizons in the process.[18]

Archeological artworks such as these are time machines, yet their way of functioning is closer to Bergson or Proust than to H. G. Wells. The user is invited to travel, but not simply up and down the shaft of time, as if encapsulated in a chronographic elevator. Instead, the traveler navigates in a much more complex realm of past-present and present-past, in which layers of time overlap and associate with each other; the conception of time is cyclical rather than linear. It is extremely important that these time machines are not automatic or remote-controlled means of (mass) transportation (like the cinema), but individual "hand-driven" vehicles. They traverse a realm only open to the active participant, who is ready to leave the customary chronological ordering of things, and the safety of his or her own socially and culturally defined observation post, heading out to explore *potential* dimensions in a conversational relationship with the work.[19]

Laboratories *cum* Playgrounds

I have refrained from speaking about an archeological "genre" of media art and resorted, instead, to the more neutral and general concept "approach." By doing so I have underlined the relatively diffuse character of the phenomenon. No manifestos have been issued, no groups of artist-archeologists formed, no principles of media-archeological aesthetics laid down. The approach itself, in spite of the growing body of work, has received little critical attention so far. There are also individual differences. In the case of artists like Jeffrey Shaw, Toshio Iwai, Lynn Hershman, and Michael Naimark the archeological approach has emerged gradually but logically from their extended struggles with certain aesthetic and ideological issues. Although the archeological dimension has been explicit in Shaw's and Iwai's work for a long time, it has became apparent only recently in Hershman's and Naimark's productions. The peep show machine

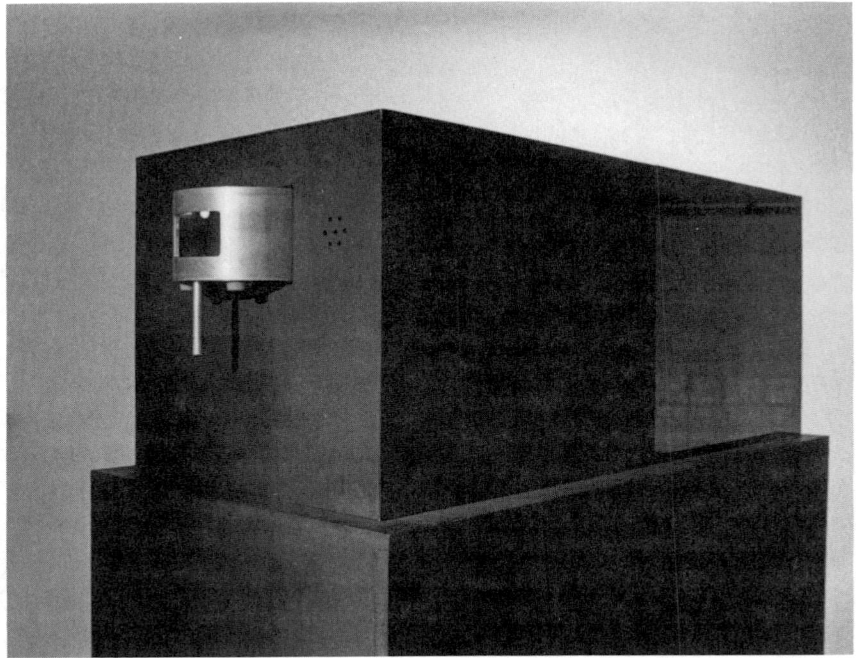

41. Lynn Hershman, *Room of One's Own* (1993), outside view. This work was produced with the collaboration of Palle Henchel and Sara Roberts. Photo Harbba Gunderstodder.

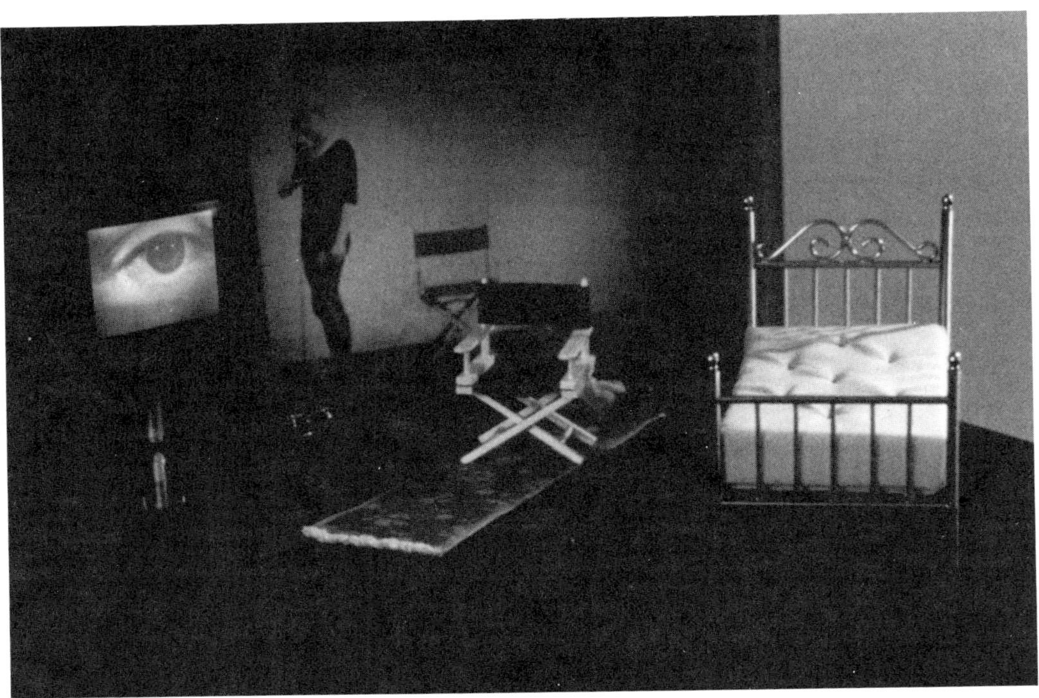

42. Lynn Hershman, *Room of One's Own* (1993), view inside the peephole. This work was produced with the collaboration of Palle Henchel and Sara Roberts. Photo Harbba Gunderstodder.

artworks that both Hershman (*A Room of One's Own,* 1992) and Naimark (*See Banff!,* 1994) have produced warrant an archeological reading of their previous works as well,[20] though it must be left to another occasion.

The ideas driving Jeffrey Shaw's art, already visible in his predigital works in the late 1960s (when he was still a member of the Eventstructure Research Group) have focused on virtual space; he has projected his questions onto realms from the illusionistic environments of baroque churches to nineteenth-century panoramas to twentieth-century satellite vision. For years Shaw has been investigating the ways in which the spectator can "break through the screen" and become an active participant, actually a "traveler" inside the image (the label Shaw uses about his art is "virtual world voyaging").[21] When virtual reality became a craze around the beginning of the 1990s, it provided a new frame of reference for Shaw's work, although his ideas had been developing along similar lines long before VR became an issue.

Throughout his career, Japanese artist Toshio Iwai has been involved in creating what a Japanese television program recently called "another evolution of the moving image."[22] Since the early 1980s his works have investigated the historical repertoire of moving image technologies, from flipbooks and zoëtropes to interactive television and videogames. Iwai constantly deconstructs, modifies, and combines audiovisual gadgets from different periods and traditions. He has described his art as a kind of response to his own dream-thoughts such as, "Wouldn't it be wonderful if a painting or an object in front of you begins to move?" or "I wish I could go into the animated space I made myself."[23] These "dreams" are really topoi, commonplace elements that have been activated many times, for thousands of years even before William Horner constructed his zoëtrope in the 1830s or Oliver Wendell Holmes described his virtual voyaging experiences made possible by the stereoscope (one successful model of which he designed) in the 1850s.[24]

Although Iwai's work can be seen as a personal vision or a private "dream history" of audiovisuality, a wider frame of reference connects it with the work of other archeologically inclined artists. This is the determination to rewrite the history of the machine culture (or parts of it) from an idiosyncratic point

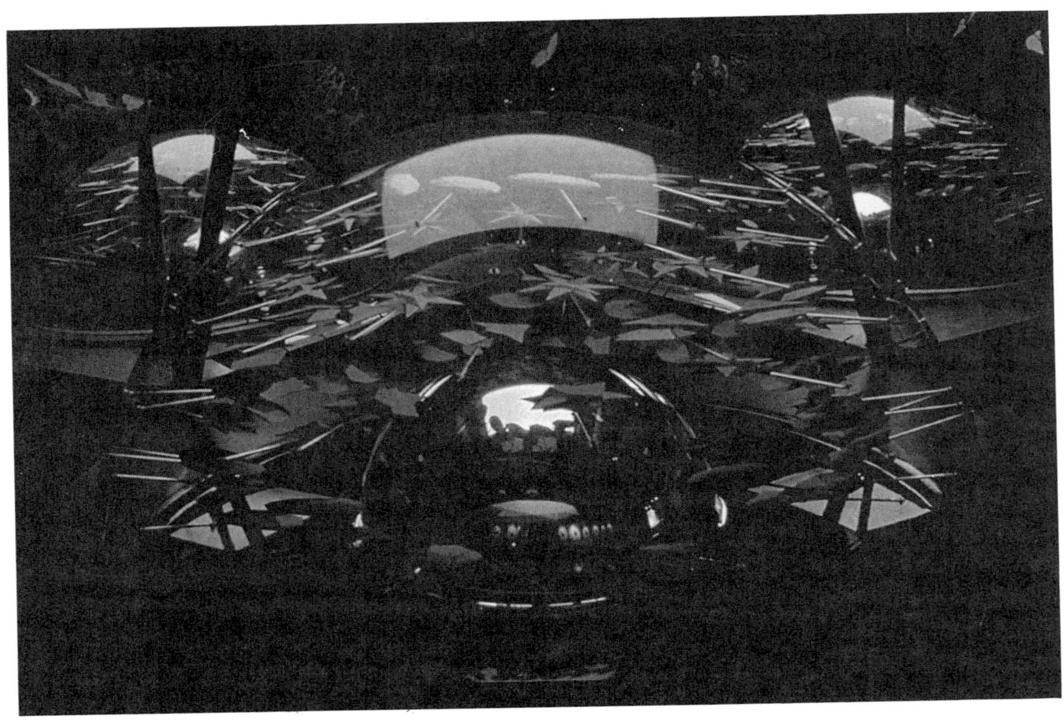

43. Toshio Iwai, *Time Stratum III* (1989). Photo courtesy the artist.

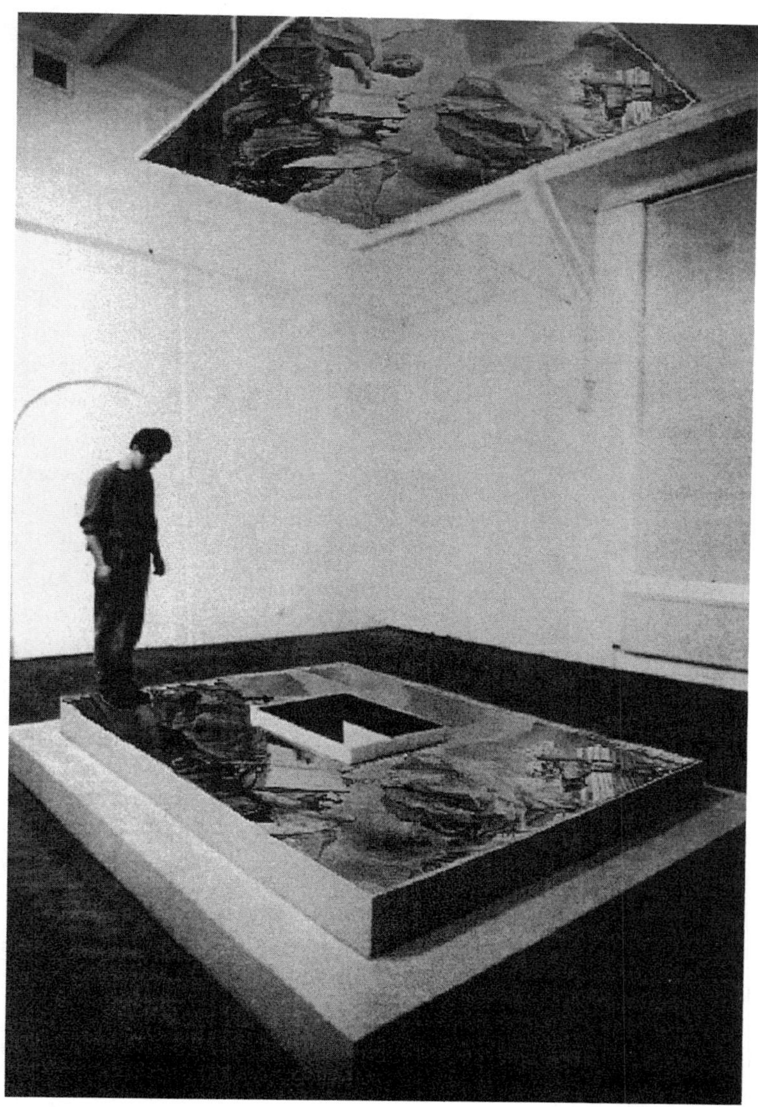

44. Jeffrey Shaw, *Heaven's Gate* (1987).

of view. History emerges in his work therefore as a series of models, or as possible histories that could have taken place, or perhaps have taken place but are hidden behind the legitimized "official" versions of history. The feminist rethinking of women's relationship to technology is another example of this. Archeological artworks have also focused attention on neglected phenomena, such as slot machines and coin-op games. These have been considered marginal, being neither useful, productive machines nor useless objects of high-culture esteem.[25]

Describing *OU,* his work in progress (an interactive "fortune telling machine," with a coin-operated speaking dummy under a glass jar), Ken Feingold has drawn a connection between old vending machines and today's interactive technology: "Vending machines . . . were one of the earliest forms of interactive objects, and still constitute the basis upon which most people form an approach to interactive artworks. That is, one 'does something' and expects 'something back', however abstract."[26] It could, indeed, be claimed that slot machines have been a significant force as an "invisible" technology, surrounding people's daily lives over the decades, accustomizing them to a one-to-one relationship with a machine ("preparing ground" for the home computer and the automatic teller machine, for example). In addition to cinema-centered film culture, a lineage can be traced from the late nineteenth-century peep show devices (often gathered in parlors or arcades) to the arcade video games of today (VR games included). Devices like Edison's kinetoscope have usually been dismissed by film historians as primitive and long since forgotten curiosities.

Artist-archeologists treat technologies of the past as cultural forms, in other words as bearers of culturally and socially assigned meanings, instead of just as technological gadgets. I have been trying to think of appropriate metaphors to cover the range of their approaches, coming up with the following: laboratories *cum* playgrounds, philosophical instruments *cum* toys, observation towers *cum* game machines. These metaphors emphasize the complex and sometimes paradoxical nature of the media-archeological undertaking. Archeological artworks establish a terrain for excavation, observation, and reflection—but they do so by defining this terrain simultaneously as a playground, children's room, or amusement arcade of a kind.[27] The playful interaction with the work has been

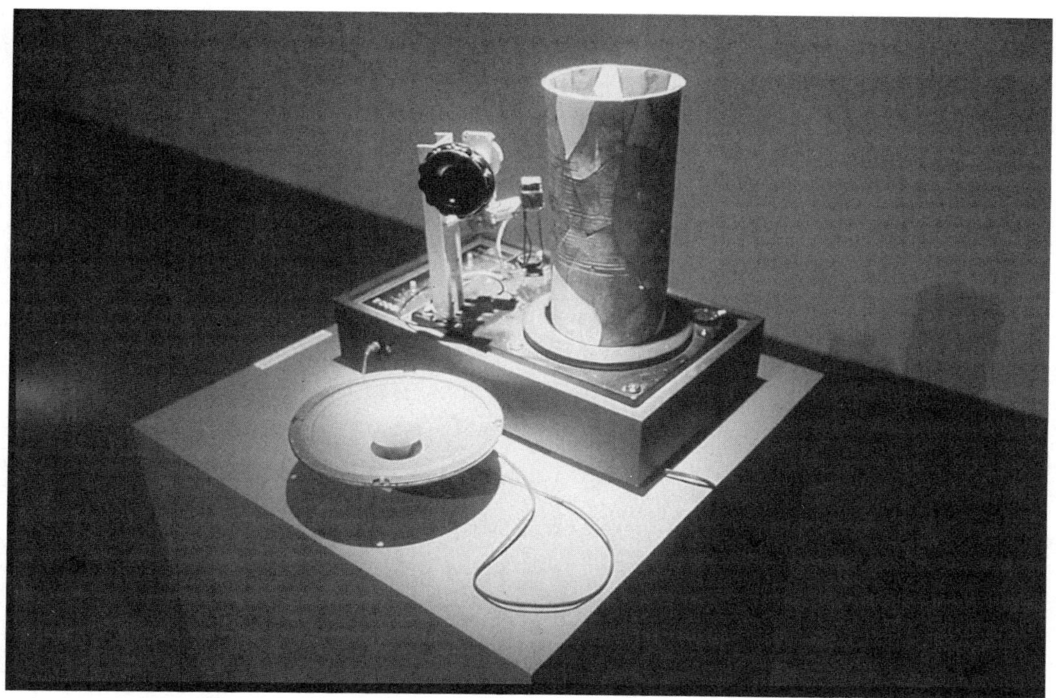

45. Paul DeMarinis, *Fragments from Jericho* (1991).

designed to entertain but in ways that should not obscure the critical project in question. It is fun to peep into the dollhouselike miniature room inside Lynn Hershman's *A Room of One's Own,* but the responses it triggers cannot be mistaken for the voyeuristic pleasures of a commercial peep show (which they are meant to deconstruct and undermine).

Although the user may not always be aware of the historical reference, she or he hopefully succeeds in reenacting the play with the subject positions and the points of view intended by the artist.[28] None of the artist-archeologists discussed here are interested in playing a self-referential postmodern hide and seek game with the user. Although none resort to the kind of didactic approach found at science centers, they all nevertheless seem to imply that media art can and should contribute to the ongoing self-assessment of the user within technoculture.

Archeological Motivations

The emergence of the archeological approach in media art in the 1990s is hardly a coincidence.[29] Technological art is seeking its identity, especially in the established fine arts world. In spite of its history, which dates back to the early twentieth century, it is often considered a newcomer still, an intruder and an alien. Its position in the computer world is not secure either; although technological art is regularly shown at events such as the Siggraph Art Show, it is frequently treated as just another ingenious application, a technological demonstration without any intrinsic aesthetic and cultural value.[30] These attitudes are more pronounced in the United States and in Japan than in Europe, although the situation is not rosy there either. Notwithstanding the select few (video artists like Bill Viola, Gary Hill, and Marie-Jo Lafontaine) who have been promoted to the rank of acclaimed "museum artists" in the eyes of the European fine arts world, even in Europe most media artists are still constrained to exhibit in the fringe, in special venues and shows that seldom draw a large general audience. On the other hand, media artists enjoy notable public funding in several European countries, and there are institutions (such as the Zentrum für Kunst und Medientechnologie in Karlsruhe) and art academies (such as the Kunsthochschule für Medien in Cologne) exclusively dedicated to promoting media art. This difference in reception may point to one of the reasons why there are relatively few media archeologists among European artists compared with American.

Media artists in the United States seem to have a more urgent need to excavate the technological past and to place their work within this cultural continuum in order to validate their practices as an art form. Their efforts are not always rewarded, however, because the roots of the archeological approach often lead the artists outside the canon of "universally accepted" high cultural forms—to technologies for selling illusions and dubitable desires, for entertaining with "canned" sounds and images, or for gambling, killing, and cleaning. Although this does not explain the media archeologist's need to prove his or her cultural viability in the art world (which does not seem to be their main urge, after all), it does offer an explanation for the appearance of the archeological approach: in media-saturated societies, classical high culture has already lost its foothold to consumerism, and popular culture gadgets like the pinball game, the Mutoscope, and the Coca-Cola vending machine (already immortalized by pop artists) have become part of a "poploristic" value system that provides present cultural forms with a legitimate foundation.

In Europe, where traditional culture and education arguably still have some strongholds, even electronic art has found its slot in the category of publicly funded "experimental art"; the acceptance of this form may represent an attempt to prove the continuing vitality of Europe's (largely lost) cultural supremacy—able to accommodate the "new" as part of the European cultural tradition. Many European media artists still "create" and "realize" themselves, with an air of solemnity reminiscent of modernist practices. They do not need to justify their position by referring to the technologies of the past.

In the context of the United States the ironic celebration of technology, apparent in the work of such diverse artists as Jeff Koons and the post-Fluxus Nam June Paik (and perhaps, to a certain extent, that of Perry Hoberman) becomes understandable.[31] The approach of the media archeologists of the

1990s, however, is somewhat different. Their affection for the debris of the machine culture seems to be intertwined with anxiety and suspicion about the role that technology actually plays in contemporary society, pushing them to investigate the processes of its becoming. A case in point, "interactive" technology is now marketed as a patent solution for almost any problem, with sales-copy promises to *enable the individual* by handing him or her the *freedom to choose* or by leading him or her to the new realm of cyberspace, where *cybersex* or *free flight in cyberspace* offer him or her experiences that are *beyond imagination*. A sharp divide is drawn between the so-called alienating technologies of the past and the new liberating technologies of the present. This readily apparent mythologization of interactivity is confronted in many media-archeological works, spanning both European and North American practices.

For a European observer like myself, a single demonstration of the commercial interactive television service in development by Time-Warner in Orlando, Florida, was startling.[32] The concept of interactivity seems to have been hijacked by corporate interests to sell more of the same in a newly designed package. The reason this may go unnoticed is the persistent failure to make the distinction between technology and cultural form that Raymond Williams emphasized.[33] A technological invention gets its "identity" only when it is placed within an institutional context. Interactive technology is not an exception. Many seem to believe that the existence of interactive gadgets, from teller machines to home computers, automatically implies a change in the human-machine relationship. However, interactive technology provides no more than a frame of opportunities, which is always filled by specific applications and ideological ideas. These may have little to do with "interactivity." Feminist artists working in the archeological idiom have understood this. Women's historical exclusion from technology and simultaneous hybridization with the machine (as telephone operators, typists, "automated" housewives, for example) is enough to generate a critical attitude toward the touted blessings of interactive technology—not to mention the widely heralded "gender-free zone" of cyberspace. These questions can be approached by way of a historical detour. The Australian cyberfeminist group VNS Matrix has chosen one such slightly different route in their project

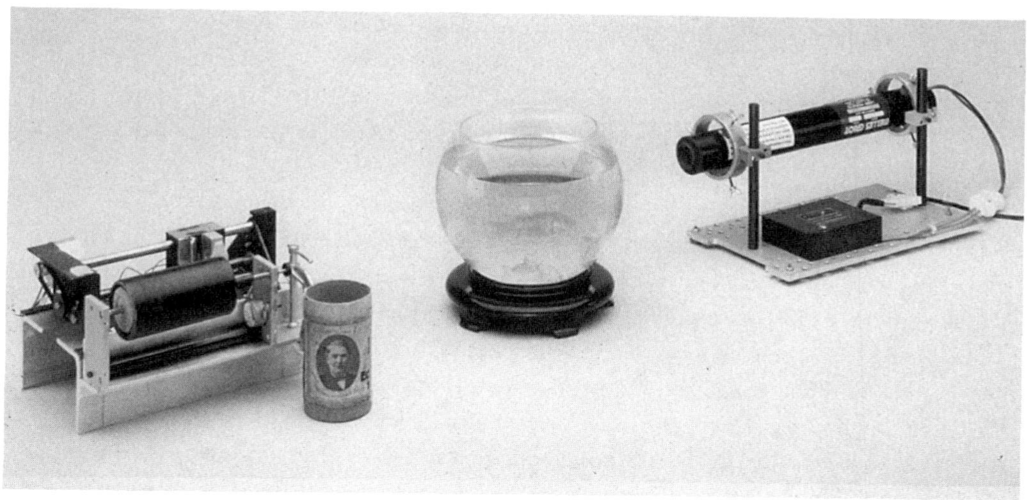

46. Paul DeMarinis, *Al and Mary Do the Waltz* (1989).

All New Gen (1993) by excavating the ideology behind one of the successful "innocent" contemporary products of patriarchal society, the Nintendo video game. By deconstructing, then reconstructing it as a cyberfeminist "game-girl" version, VNS Matrix contributes a valuable artifact to the archeological endeavor.

Against this background, the activity of the artists working in the archeological idiom becomes extremely important. The artist-archeologists maintain a constant dialog (or rather a polylog) about the nature of technoculture, unearthing its forgotten preforms and scrutinizing the differences between technologies and cultural forms. By moving constantly between the past and present they contribute to the formulation of new and hopefully more versatile interfaces and applications. Last but not least, they combat the mythologization of phenomena such as interactivity by exposing the ways in which great promises are used as just another marketing formula, skillfully camouflaged behind the chrome facade of the brave new world of technoculture.

Notes

1. The Artist-in-Residence program at the Exploratorium in San Francisco has been by far the most significant of these. Since the inception of the program in 1974, four to six artists per year have produced work, much of which has been shown extensively. The list of artists includes Jim Pomeroy, Ed Tannenbaum, Paul DeMarinis, Skip Sweeney, and Toshio Iwai. Mike Naimark produced his *Golden Gate* movie map there (1987). La Villette in Paris has commissioned Jeffrey Shaw's *Inventer la terre* (1986) and other works. It has also maintained an Atelier Virtuel as a showcase for contemporary media art.

2. In this article I use the word *machine* for both mechanical and digital gadgets, fully conscious about the discussion of whether a computer can be considered a machine or something else. This distinction is reflected in the title of a television series about early cinema, *The Last Machine,* produced by John Wyver/Illuminations in England (1994). It is indeed quite intriguing that artists working with digital technologies now show so much interest in *mechanical* technology (symbolized by one of the pieces in Toshio Iwai's *Man-Machine TV* installation series, in which the participants use a mechanical crank to

animate a computer-generated robot on a screen!). For discussions about the definitions of the machine, see Bruce Mazlish, *The Fourth Discontinuity: The Co-evolution of Humans and Machines* (New Haven and London: Yale University Press, 1993); and Raymond Kurzweil, *The Age of Intelligent Machines* (Cambridge, Mass.: MIT Press, 1990).

3. Feingold has emphasized the personal memories underlying his work: "I have undertaken a search for my childhood TV memories, a kind of archaeology of those images and sounds that I remember, or see now, as having been formative in my understanding of what was going on in the world," in *Iterations: The New Image,* edited by Timothy Druckrey (New York and Cambridge, Mass.: International Center of Photography and MIT Press, 1993), 164.

4. Equally "archeological," but somewhat outside my concern here, is George Legrady's installation *An Anecdoted Archive from the Cold War* (1993), also released as a CD-ROM. Legrady, a Hungarian living in the United States, uses hypermedia to reconstruct his family's memories about pre-1956 Hungary and the ensuing expatriation, combining private and official history from a wealth of documents from communist literature and posters to drawings by family members.

5. For an extensive treatment of this issue from the fifteenth century to the late nineteenth century, see Martin Kemp, *The Science of Art: Optical Themes in Western Art from Brunelleschi to Seurat* (New Haven and London: Yale University Press, 1990); see also Jonathan Crary, *Techniques of the Observer: On Vision and Modernity in the Nineteenth Century* (Cambridge, Mass.: MIT Press, 1991).

6. Frank Popper, *Art, action et participation: L'artiste et la créativité aujourd'hui* (Paris: Éditions Klincksieck, 1980; reprint 1985); Jack Burnham's *Beyond Modern Sculpture* (New York: George Braziller, 1968); see also the extremely rich catalogue, *Electra: L'électricité et l'électronique dans l'art au XXe siècle* (Paris: MAM Musée d'Art Moderne de la Ville de Paris, 1984).

7. C. P. Snow, *The Two Cultures and a Second Look* (Cambridge: Cambridge University Press, 1979, Part One, 1959). See also Bill Buxton, "Snow's Two Cultures Revisited: Perspectives on Human-Computer Interface Design," *Cyberarts: Exploring Science and Technology,* ed. Linda Jacobson (San Francisco: Miller Freeman, 1992), 24–31.

8. Ben F. Laposky, "Oscillons: Electronic Abstractions," in *Kinetic Art: Theory and Practice: Selections from the Journal Leonardo,* edited by Frank J. Malina (New York: Dover Publications, 1974), 150.

9. Myron W. Krueger, *Artificial Reality II* (Reading, Mass.: Addison-Wesley, 1991).

10. See the wonderful exhibition catalog *Eigenwelt der Apparatewelt: Pioneers of Electronic Art,* ed. David Dunn (Linz: Ars Electronica, 1992). The exhibition was curated by Woody and Steina Vasulka.

11. Woody and Steina Vasulka's art could also be called "archeological" in the sense that it often incorporates obsolete electronic gadgets or their parts. The emphasis is, however, primarily on the technological and formal rather than on the historical and cultural issues. Yet Woody's recent installation *Brotherhood* (1994), which appropriates old military technology and also uses it as subject matter, falls definitively within my definition of the archeological tendency. On the other hand, the Vasulkas are great archeologists and collectors of electronic technology. The exhibition of early video synthesizers, most of which worked and could be manipulated by the audience, curated by them for Ars Electronica 1992 in Linz, Austria, was a media-archeological event of the highest order.

12. About the idea of making a distinction between technology and cultural forms, see Raymond Williams, *Television: Technology and Cultural Form* (Glasgow: Fontana, 1974; reprint 1979). About "discursive inventions," see my "From Kaleidoscomaniac to Cybernerd: Towards an Archeology of the Media," *ISEA '94 Catalogue,* ed. Minna Tarkka (Helsinki: The University of Art and Design, 1994), 130–135.

13. According to the videotape *HDTV & The Quest for Virtual Reality* (*ACM Siggraph Video Review,* issue 60), the honor for coining the concept "interface" in relation to the human-computer relationship goes, however, to George W. Hoover, a flight simulator developer in the United States Air force. Hoover pioneered the merging of the real-time imaging capabilities of the digital computer with the previously largely mechanical flight simulators (or "trainers") in the 1960s. Again, "interface" is linked to the issue of ideology from the outset.

14. For a more detailed analysis of these works, see my "Commentaries on Metacommentaries on Interactivity," *Cultural Diversity in the Global Village: The Third International Symposium on Electronic Art,* ed. Alessio Cavallaro et al. (Sydney: Australian Network for Art and Technology, 1992), 93–98. A revised version of this essay, "Penetrating into the Virtual World: Ways of Communicating with Interactive Art," was published in *The European Media Art Festival 1993* catalog, ed. Eckhard Diesing and Ralf Sausmikat (Osnabrück: Experimentalfilm Workshop e.V., 1993), 262–272.

15. An interesting parallel with the archeological projects by these artists was provided by the exhibition *Mechanical Brides: Women and Machines from Home to Office,* curated by Ellen Lupton and exhibited at the Cooper-Hewitt National Museum of Design in New York (August 1993 to January 1994). The well-researched and documented catalog by the same name, also by Lupton (New York: Princeton Architectural Press, 1993) draws from cultural studies and feminist theory to analyze women's historical relationship with technologies, from the washing machine and the vacuum cleaner to the telephone and the typewriter. Jill Scott's *Machinedreams* was recently shown at the Kunst- und Ausstellungshalle der Bundesrepublik Deutschland, Bonn, Germany, from September to November 1994. About Heidi Kumao's art, see Lynn Love, "Hidden Mechanisms: An Interview with Heidi Kumao," *Afterimage* 21:7 (February 1994), 6–9.

16. An exception could be the art of Mark Pauline and the Survival Research Laboratories, although the archeological references in their destructive machine performances are usually metaphoric rather than precisely addressed. In addition, the work of SRL functions as an extreme testing ground for heavy duty technology, which has raised doubts about the origins of the expensive technological arsenal (sometimes rumored to have been "found" or stolen) used by this subversive artist group.

17. Another artwork using the peep show device is Catherine Richards's *Virtual Body* (1994). Peering into an eyepiece on top of a beautifully finished wooden case (or column), one "enters" an illusionistic miniature rococo salon. Sticking one's hand into this "magic box" through a hole on one side triggers interactive changes in the image space, which "collapses" and causes a feeling of disorientation and weightlessness, a kind of loss of the (physical) body. Related ideas about illusionistic image environments and their "dematerializing" effect on the body were earlier investigated by Jeffrey Shaw in

his installation *Heaven's Gate* (1987), in which Shaw made an intriguing connection between baroque *sotto in sù* painting and satellite images shot from orbit.

18. There are many other aspects we could consider. It is very significant that both Naimark's and Shaw's works evoke the virtual reality interface. Naimark locates its historical sources in the peep show machine and the stereoscope. Shaw's piece reenacts the situation of Ivan Sutherland's first head-mounted display ("The Sword of Damocles") from the late 1960s by superimposing the virtual worlds on the real world in the user's field of vision. For a more detailed study of Shaw's work, see my "Virtual Voyaging in the Landscape of Doubt," *Media Passage: InterCommunication '93: Agnes Hegedüs, Matt Mullican, Jeffrey Shaw,* ed. Akihiko Yoshimura and NTT Publishing Co. (Tokyo: NTT, 1993), 42–49.

19. Rebecca Cummins's *To Fall Standing* (1993) is another time-traveling installation that raises issues about looking and power. Its central element is an old gun, reminiscent of Etienne-Jules Marey's chronophotographic gun (1882), which was a hybrid between a means of mass destruction, a scientific instrument, and a forerunner of the motion picture camera. By pointing the gun at another visitor and pulling the trigger, the participant freezes the other person's video image, seen in a series of monitors and captured by a miniature video camera inside the gun. The video image has been run through a Fairlight computer and turned into a continuous series of still images reminiscent of Marey's chronophotographic series of human beings or animals in motion. The outer form of the installation adds to the complexity by evoking a fairground shooting arcade. Lynn Hershman's interactive installation project, *America's Finest* (1994), uses a gun provided with a video camera as the interface device as well.

20. Naimark's archeological interests have become clear also from his writings, for example in *Elements of Realspace Imaging: Technical Report* (San Francisco: Apple Multimedia Lab, 1991). Hershman has given an archeological description of *A Room of One's Own,* realized with Palle Henckel and Sara Roberts, in *Iterations: The New Image,* ed. Druckrey, 150–152, saying that it "references early 'peep shows'" (150). The historical facts in her account are not completely accurate; for example the claim that "Marey perfected a gun that substituted film for bullets" should read "a series of photographic plates" instead of film, which only became available in 1888. The claim that Edison's

"kinetograph" was "an alliance between the phonograph and the photograph" takes a short cut. "Kinetograph" was the motion picture camera invented by W. K. L. Dickson for Edison, while the viewing device was called the "kinetoscope." This was a silent device. An "alliance between the phonograph and the photograph" also came into existence, but it was called the "kinetophonograph" (See W. K. L. Dickson and Antonia Dickson, *History of the Kinetograph, Kinetoscope & Kinetophonograph* (Salem, New Hampshire: Ayer, 1895; reprint 1984). But the idea seems more important to the artist than the accuracy of the historical background.

21. Shaw's entire artistic output can be seen as a far-ranging and open-ended conversation with different mechanisms of representation and simulation, ranging from the visual machineries or "scopic regimes" (a term coined by Martin Jay) of the Renaissance and the baroque to the different "presences" of the electronic image (Martin Jay, "Scopic Regimes of Modernity," *Vision and Visuality,* ed. Hal Foster [Seattle: Bay Press, 1988], 2–23). See also Martin Jay, *Downcast Eyes: Denigration of Vision in 20th Century French Thought* (Berkeley: University of California Press, 1993). Shaw places spectators in highly complex perceptual situations and makes them investigate their own relationship to them. These perceptual mechanisms do not work simply as formal devices. Rather, they carry inscriptions of the discursive traditions that have informed and molded them historically. See my "Virtual Voyaging in the Landscape of Doubt."

22. In *What's Next,* TV-Asahi, 10 March 1991. For discussions of Iwai's work, see *Toshio Iwai,* exhibition catalog, ed. Päivi Talasmaa (Espoo, Finland, and Karlsruhe: Galleria OTSO & ZKM, Medienmuseum Karlsruhe, 1994), with essays by Erkki Huhtamo, Yoshitomo Morioka, and Toshio Iwai.

23. "In Pursuit of a New, Natural Relationship between Image, Machine and Man," statement in the poster for Iwai's exhibition *Machine for Trinity,* Mitsubishi-Jisho Artium Gallery in Japan, 1990.

24. About the topos theory applied to the study of media culture, see my articles, "From Kaleidoscomaniac to Cybernerd," and "Encapsulated Bodies in Motion: Simulators and the Quest for Total Immersion," *Critical Issues in Electronic Media,* ed. Simon Penny (Albany: SUNY Press, 1995).

25. The sociologist Yves Hersant has written aptly: "As far as mechanism is concerned, the slot-machine appears to be an anti-motor. As far as economics, it is anti-productive. It wastes any effort and investment put into it. Give it something and it will do nothing with it, using all its technical resources to achieve nothing." (Introduction to Jean-Claude Baudot, *ARCADIA: Slot Machines of Europe and America,* trans. Anthony Carter (Tunbridge Wells, Kent, England: Costello, 1988). Typically, most accounts of the history of the slot machine are picture books meant for collectors. See also Bill Kurtz, *Slot Machines and Coin-op Games: A Collector's Guide to One-armed Bandits and Amusement Machines* (London: The Apple Press, 1991). The most useful background study remains Sigfried Giedion's "contribution to anonymous history," *Mechanization Takes Command* (New York: W. W. Norton, 1948; reprint 1969).

26. Ken Feingold: "*OU:* Notes on a Work in Progress," April 1993 (unprinted leaflet), 3–4. Feingold analyzes interestingly the historical position of the fortune-telling machine: "Fortune-telling *machines* came into existence (in western cultures) with mechanization in general. Like many other machines, they served not to make life or work easier, but to circumvent direct encounter and interaction between seller and buyer. In such a culture, where divination is generally discredited and marginalized, and those who perform it are often objects of scorn . . . the fortune-telling machine allowed a kind of pseudo-divinatory activity to take place, in the guise of purchasing entertainment" (3).

27. I prefer these to the metaphor of the theater used by Brenda Laurel to describe the human-computer interaction in her *Computers as Theater* (Reading, Mass.: Addison-Wesley, 1991). The theater metaphor remains forced, reflecting perhaps Laurel's own previous (insider's) experience as an actress, but not that of any regular theater audiences. The great majority of theatrical experiences in the western tradition (and even in Japanese Kabuki, for example) are basically noninteractive, drawing a clear demarcation line between the (receiving) audience and the (transmitting) actors. The situations in which the audience has been invited to step up on the stage and join in have been exceptions, even though this certainly happens in collective rituals (one of drama's historical roots) and in children's games. The interfacing of the human and the computer is fundamentally different. It involves an active exchange between an individual and a cybernetic system (and with another individual, in network communication). Even cinema as an audiovisual experience generator would have been a more appropriate metaphor, although the

relationship is problematically reduced to the "mental" interaction spurred by the functioning of the cinematic apparatus.

28. One can make a comparision with the Structural Film Movement of the 1960s and the early 1970s. Many structural filmmakers wanted to return to Lumière, to investigate the strategies of the early cinema in their search for a kind of cinema other than the alienating commercial cinema. Their strategies were theoretical and conceptual, negating the pleasure principles of the latter. Filmmakers Ken Jacobs, Malcolm LeGrice, Al Razutis, and others resorted to recycling found footage from early silent films; others (not always strictly structural filmmakers) such as Stan Brakhage, Michael Snow, and Werner Nekes (and also, but partly within the structure of the commercial cinema, Jean-Luc Godard) in a way purported to recreate the history of the cinema from their own subjective points of view. In the 1980s both Nekes (*Was geschah zwichen den Bildern? or Film before Film*) and Godard (*Histoire(s) du Cinéma*) elaborated their position in explicitly archeological works. See Noël Burch, "Primitivism and the Avant-Gardes: A Dialectical Approach," *Narrative, Apparatus, Ideology: A Film Theory Reader,* ed. Philip Rosen (New York: Columbia University Press, 1986), 483–506. For a general discussion, see Bart Testa, *Back and Forth: Early Cinema and the Avant-Garde* (Toronto: Art Gallery of Ontario, 1992).

29. There is also a parallel with the recent historizing tendency in media studies, evident in the work of Anne Friedberg, Avital Ronell, Tom Gunning, William Boddy, Lynn Spiegel, Susan Douglas, Carolyn Marvin, Siegfried Zielinski, Friedrich Kittler, and others. See my "From Kaleidoscomaniac to Cybernerd."

30. The way artists were treated at the Edge show at Siggraph 94 in Orlando, Florida, sparked an intense revolt among the art community and gave birth to the Sigart list, a debate on the Internet about the problems faced by media artists. About the problematic position of technological art between the art world and the computer world, see my "It Is Interactive—but Is It Art?," *Computer Graphics Visual Proceedings,* Annual Conference Series, ed. Thomas E. Linehan (New York: Association for Computer Machinery, 1993), 133–135.

31. Japanese media artists, especially from the younger generations, share the celebration of technology while they seem to lack irony. Many young media artists have been inspired recently by video games, but instead of scrutinizing their cultural and ideological underpinnings, they create their own artistic and playful games based on the old ones in the spirit of joyful recycling. The work of Shigeyuki Handa, which has been inspired by *Pacman, Space Invaders,* and other classic video games, is a good example. In his *Game Center Self Conclusion* (1994) Handa captures the player's face with a video camera; the face thus becomes an element of his game simulations. Toshio Iwai's culturally and historically informed archeological art is rather an exception, even though the practice of referring to old tech is quite common. But even Iwai's art is about discovery more than about rediscovery; he often refers to technologies that did *not* exist in the Japanese context in the past. The interpretation of Iwai's work in the West is, of course, entirely different.

32. I am referring to a videotape demonstration by Robert Zitter (*Home Box Office*) at the Imagina conference at Monte Carlo in February 1994. The system is called Full Service Network (FSN).

33. Williams, *Television: Technology and Cultural Form.*

Artists´ Statements

Objects of Ritual

Will Bauer and Steve Gibson

Objects of Ritual is about how humanity builds virtual worlds—how human society is a large-scale virtual simulation, with billions of information-processing elements (humans) each participating in its construction and maintenance. An underlying thesis of this artwork is that for humans, everything is a virtual world—in a sense, virtual reality is not new but, rather, unavoidable. It is never possible to directly perceive what is "really" outside us; all our senses are abstractions constructed out of incoming data flows. Patterns recognized at low levels are linked and correlated to other patterns to make still more abstract structures until finally, at a conscious level, we see a bouquet of daisies or hear twelfth-century choral music. Human rituals and the physical objects associated with them become software gateways to the manipulation of "societal cyberspace," an abstract dataspace created by human consciousness and interpersonal communication. In *Objects of Ritual,* four participants conduct an electronic ritual during which they build a virtual world of their own, sustain it by their presence and communication, and cause it to collapse upon itself when they leave.

Our relationship to technology is a simultaneous duality: we embrace it because it makes the impossible possible; we hate it because, all too often, it is brittle and recalcitrant. We view technology as a way of extending mind *and* body. Part of our aesthetic and hence part of our reason for choosing to create a piece like *Objects of Ritual* involves the idea that you can "rediscover your body as machine." The human body is an amazing piece of high technology, far more complex than anything humans have yet learned to build. As such, we believe it should be *the* focal point when talking about technologic art. When it comes to deciding what is "real," man-made digital "mass storage" external memories are replacing the internal firmware memories contained in our brains. For the first time in human history, it is now possible to construct internally self-consistent realities-as-objects on a widespread basis. It will soon be possible to manufacture these reality objects as easily as industrial age factories turned out pins or threaded screws. What are the social implications of such a capability? This is one of the themes we wanted to raise, at least subliminally, as part of *Objects of Ritual* and one of the reasons we chose to create a piece about society and VR/cyberspace.

Objects of Ritual was created as an experiment in immersive group experience of real-time media control. We wanted to create a piece that blurred or erased some conventional boundaries between performer and audience. To this end, we decided to "just say no to helmets" and to eschew the use of conventional VR tracking and visualization equipment. The replacement for these tools was a system originally developed by one of us (Bauer) that is now available commercially. GAMS (the Gesture And Media System) allows performers to control and manipulate electronic media in real time. It is a wireless three-dimensional tracking system that allows the tracking of up to four "wands" over large areas (up to about 8 meters by 8 meters [26 feet by 26 feet], or under some conditions even larger). Three-dimensional wand position and velocity are monitored up to thirty times per second and can be used to control any MIDI-addressable media (such as music samplers, slide controllers, and lighting controllers) or can be passed along to other computer platforms (such as SGI or NeXT machines) for their specific use.

The other primary technology used to integrate media in *Objects of Ritual* is the Digital Performer software for the Macintosh. In *Objects of Ritual,* Digital Performer is used as a master interface, managing data between MIDI instruments, GAMS, the narration (stored on hard disk), slides, a NeXT computer, and an SGI computer. Needless to say, this goes beyond the original intentions of the software. We created special remote controls to allow GAMS to turn sequences on and off. In addition we used MIDI controllers to control the fades and morphs on the SGI in real time. The slides are advanced in real time by certain MIDI notes that send pulses via a specially designed interface. In essence the entire performance is accessed and run in real time via one medium-sized file on a Macintosh computer.

Objects of Ritual was intended as an immersive VR group experience and was designed from the beginning to require the participation of a minimum of three people. Our aesthetic was that of art-as-information, that is, that the experience of art is due to the extraction of information by the person experiencing the work. Of course, this extraction or decompression of information is highly dependent on the local context of the person—his or her experience, knowledge, and personality—as well as on the form of the information being presented. For this reason, structure was important to us, and we created a number of broad design principles that shaped the overall form of the piece.

We chose a theatrical or perhaps operatic structure within which the participants would control the pacing of events and be able to directly control some of the media at certain times during the piece. We also desired limited interactivity. The reasoning was that if we did not provide fairly simple interactive structures, people not practiced in the manipulation of such environments would drown in a sea of possibilities. A guiding analogy was to ask ourselves what would happen if a nonmusician was given a violin and then, moments later, invited to give a performance; clearly a social situation best avoided at all costs.

The story line of *Objects of Ritual* is linear, split into two main sections. The first part is passive. Participants are solely observers, occupying only the physical space while the virtual is, to them, unreachable. During the second portion of the piece, participants begin to control and create within the virtual space. By the end of the piece, participants are given total control over the virtual space, which is overlaid into the seven-meter square of physical space via image projections on four screens surrounding the space and sounds coming from spatially distributed speakers. Participants are able to observe and manipulate virtual objects in three dimensions as they moves through the physical space. At the piece's end, the participants must leave. This causes the virtual world they have so carefully built up to collapse upon itself, since it is the presence of, and communication between, the participants that sustains the world.

What are the implications of digital technology and virtual reality for narrative form? Can we speak of organic approaches to structure in the age of poststructuralism? What are the relationships between theatrical narrative, musical narrative, and technology?

Objects of Ritual is a direct attempt to articulate an organic structure based on the technological imperatives of the employed media. Thus the narrative of *Objects of Ritual* is based on the idea of the virtualization of reality, with the obvious musical/visual corollary being the virtualization of the employed sound objects and visuals. The pure soprano sample of the opening becomes increasingly artificial, the visual elements morph into distantly related objects, the general musical world becomes more and more artificial-sounding.

Objects of Ritual thus ultimately argues for/against the use of virtual reality as a tool for distortion rather than mimesis. It questions the commonly held assumption that virtual reality is primarily useful as an imitator of the real and a simulator of certain realities. Virtual reality distorts, manipulates, and it tells lies—like all art.

Is it possible to talk about expressivity in connection with virtual reality and digital technology? Or is the public perception that technological art is an inevitably cold medium becoming a self-fulfilling prophecy?

Objects of Ritual unabashedly pays homage to the expressivity of opera as media. How does opera turn the ridiculous into the sublime? In essence it manipulates our feelings by means of expressive musical devices, combined with extreme emotional situations. *Objects of Ritual* uses musical devices similarly, also with the aim of creating an expressive experience. The difference is that here we have the cool medium of virtual reality, with limited human interface. The effective result is a creative tension between media and materials, a clash between form and feeling, and a merging of the virtual brain with the virtual body.

Archeology of a Mother Tongue

Toni Dove and Michael Mackenzie

Toni Dove

The development of immersive and interactive technologies reflects some profound changes in cultural relationships to the constructs of performer, space, and audience. Walter Benjamin's concept of film as a redemptive medium, a technology that gave back a coherent image of the fragmented experience of an era in transition and replaced experience that had been depleted by the mechanizations of the industrial revolution, offers an interesting point of departure for looking at the current escalations of the technologies of spectacle.

Virtual reality, rather than existing as a singular medium, has become a mythic and metaphoric cultural repository for the anxieties of loss and change brought about by technological transformation. An obsessive concern with presence and community saturates environments (CD-ROM games, virtual reality theme parks, World Wide Web sites, MOOs and MUDs, Imax theaters) that paradoxically move an isolated viewer away from communal audience experience in physical space and into a focused tunnel vision relationship to technological apparatus. The content of these environments frequently addresses the loss of physical space and human contact with a rhetoric of replacement. Narrative strategies shift away from character- and plot-driven stories to experiences that escalate physical and visual intensities. The cut of the film gives way to the continuum of interactive and immersive experience. An interesting interim example is the rock video—a form driven by visual and physical sensation that is also, not accidentally, an example of the disappearance of the advertisement as a discrete entity—the product and the advertisement fuse and become one. This is a form located somewhere at the intersection of the advertisement, the movie, the concert, and the theme park ride. If this tunnel vision into technology is designed on some level to provide a more effective means of social control, then it is an interesting project for artists interested in these developing media to attempt intervention into the formation of this syntax. The Virtual Art and Environments Project at the Banff Centre for the Arts was a rare opportunity for a group of artists to do just that.

Archeology of a Mother Tongue started as a conversation and was conceived of initially as a multimedia theater piece. Michael Mackenzie and I talked about the city as a maternal body and the body as a map. The murder mystery in the narrative included an autopsy that became a conceptual bridge between the physical body and the city. Forensic evidence from a corpse led back to the city (dirt under fingernails, patterns of behavior) and described a life through layers of mapped evidence—trails. We were both interested in constructing different approaches to a narrative engine or to the vehicle that moves a story through time. The use of interactivity in the context of virtual reality seemed an interesting ground for experimenting in this area. The map, the wireframe simulation, the wireframe as architecture and as the body, became a place to start building a multidimensional interactive environment. We made a decision to design the environment for an audience with a single interactive performer and to avoid the more hermetic (tunnel vision) experience of the enclosing head-mounted display. We were interested in using the technology as a narrative metaphor with critical possibilities to examine some of the social and cultural ramifications of a new and powerful medium. If artists can have an impact on the developing vocabulary of interactive media it seems important to look beyond the formal and structural possibilities to the economic and political context of their use.

I found that one of the most interesting aspects of the work process on this project was the division of labor and collaboration that occurred within

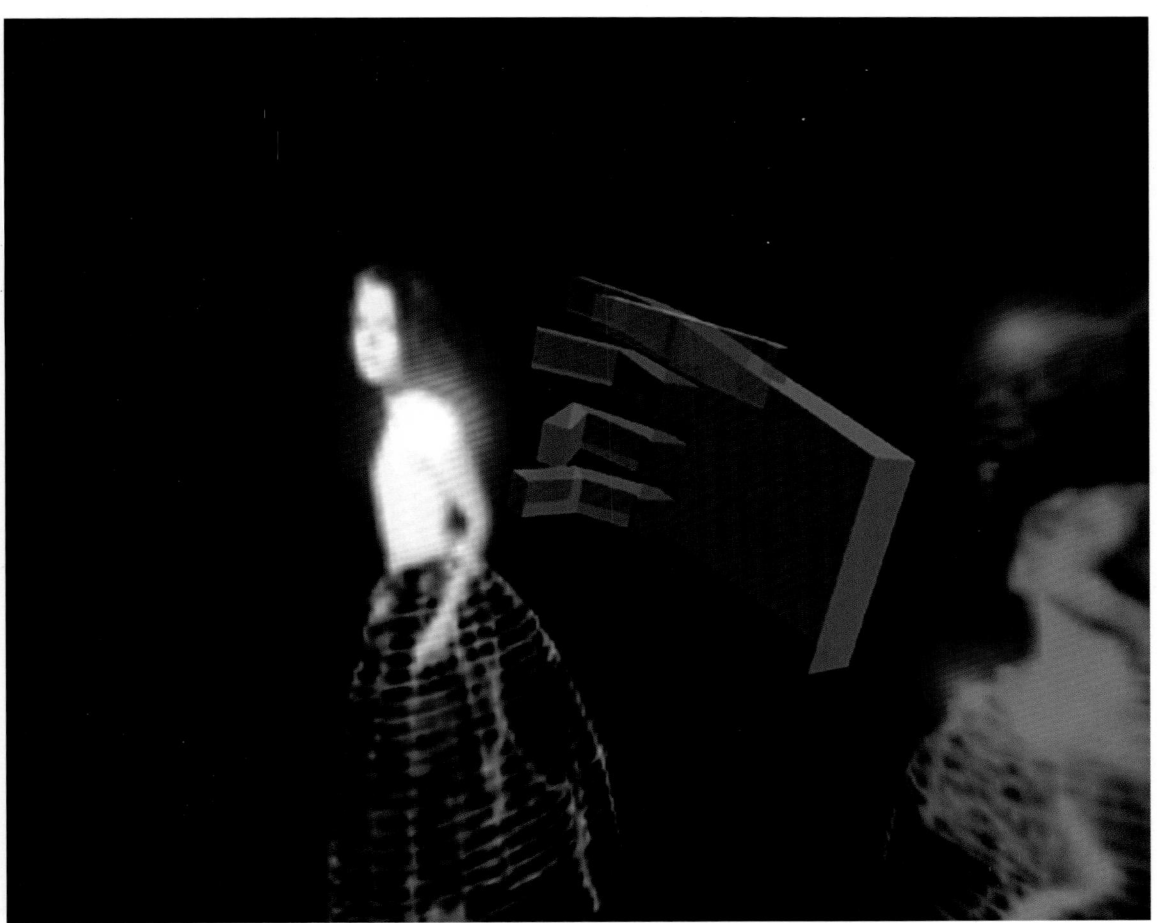

Plate 1. Toni Dove and Michael Mackenzie, *Archeology of a Mother Tongue* (1993), frame grab from the prolog of the piece. Digital image courtesy the artists.

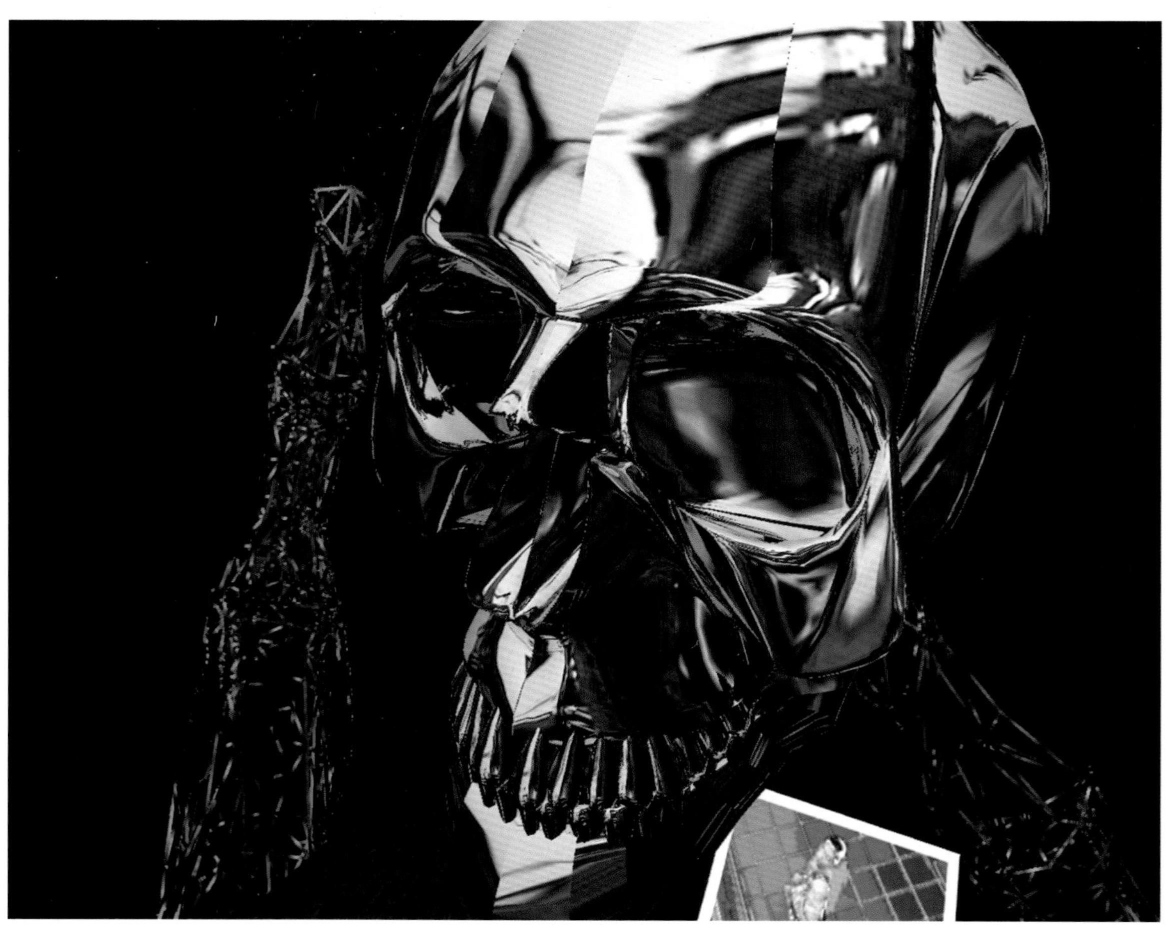

Plate 2. Toni Dove and Michael Mackenzie, *Archeology of a Mother Tongue* (1993), frame grab from one of three narrative "envelopes" or virtual environments. Digital image courtesy the artists.

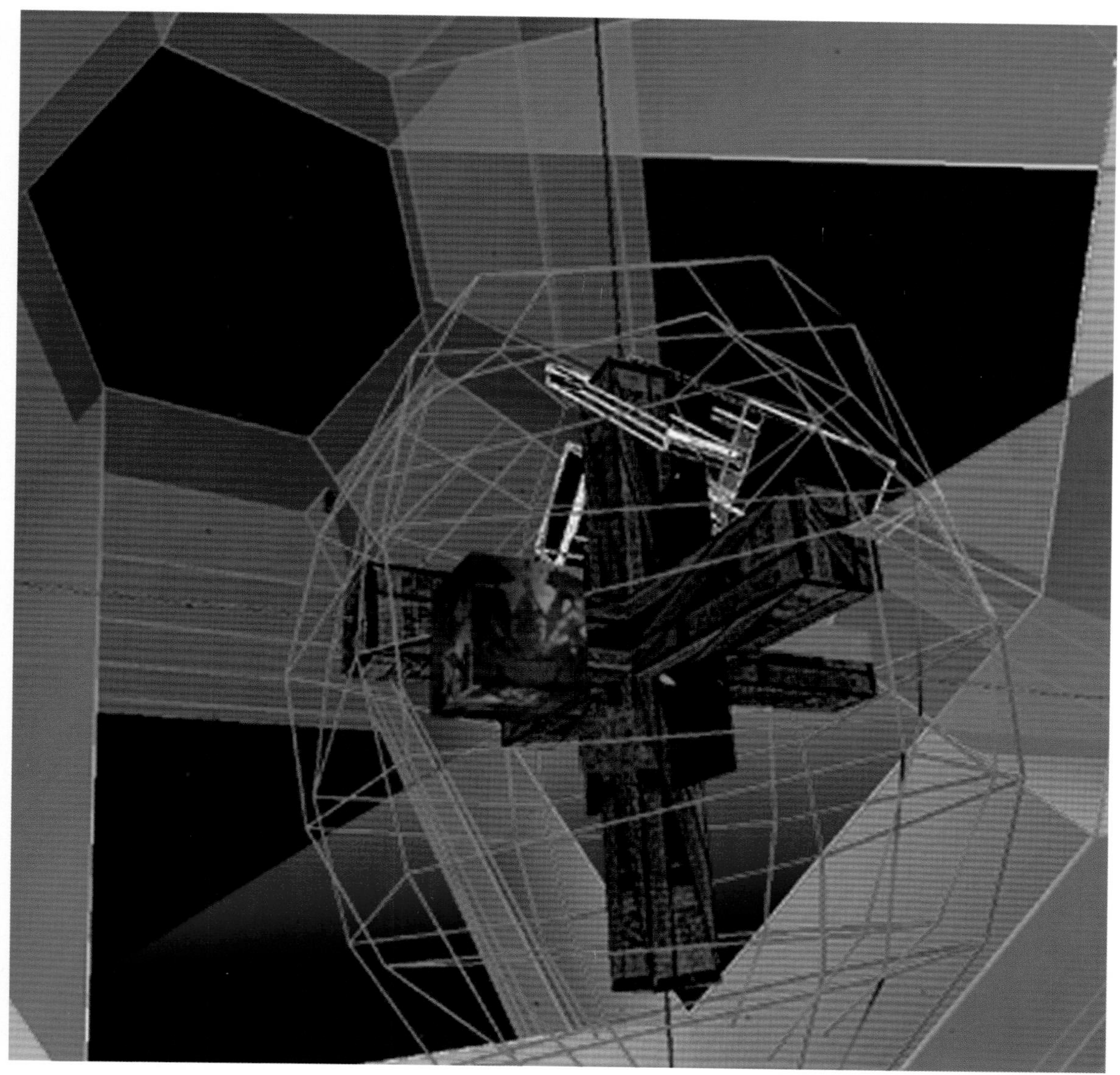

Plate 12. Marcos Novak, *Dancing with the Virtual Dervish: Worlds in Progress* (1994), frame grab of one of four virtual worlds. This one shows the four-dimensional world. Digital image courtesy the artist.

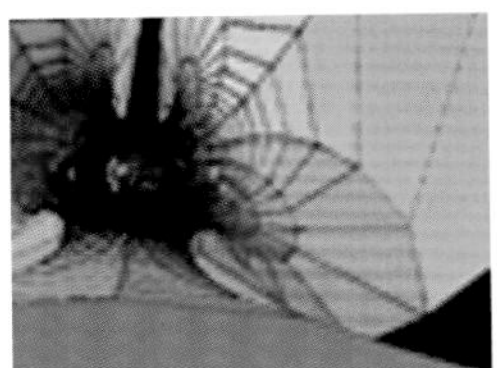
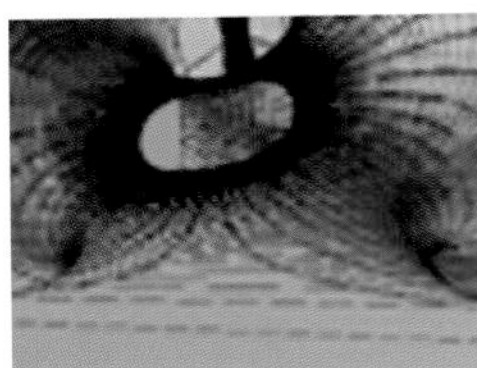
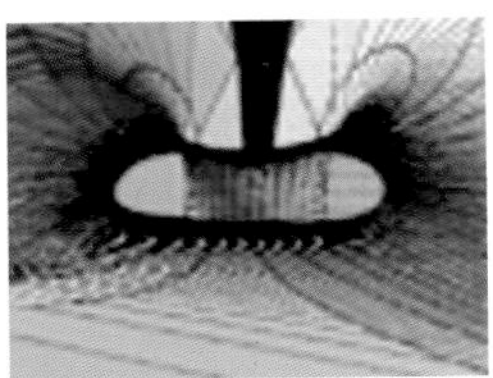
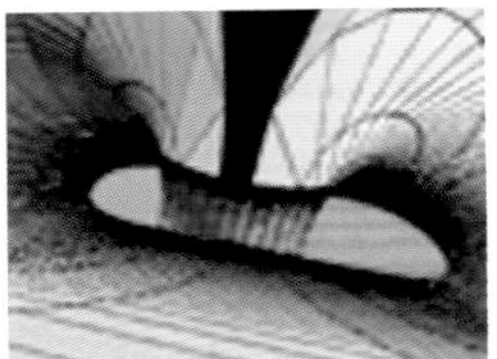
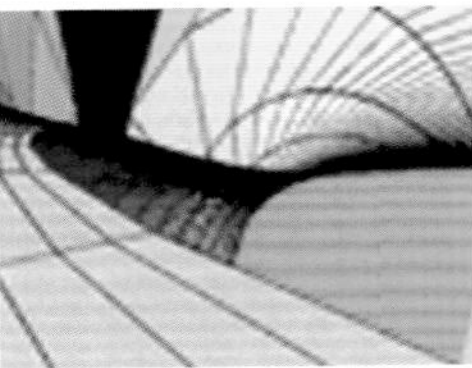
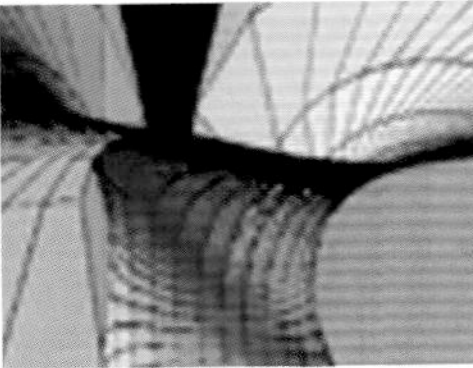
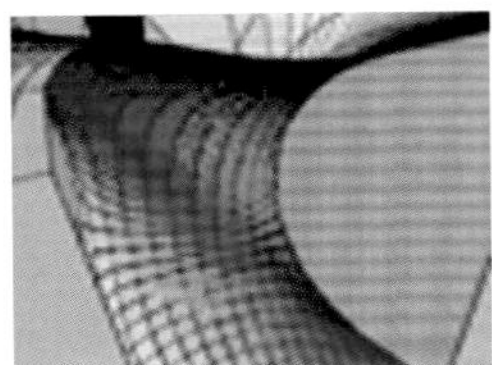
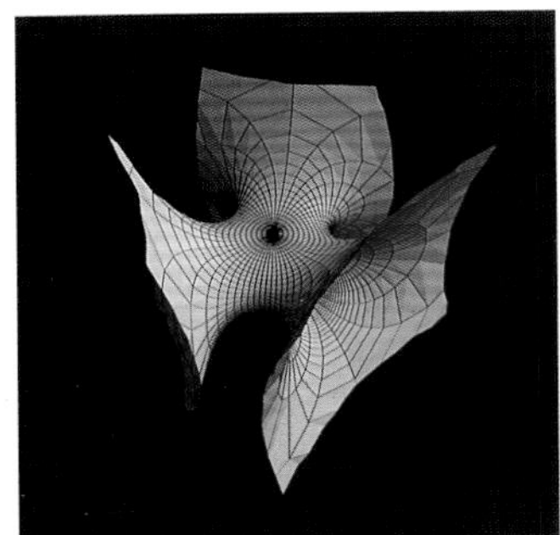
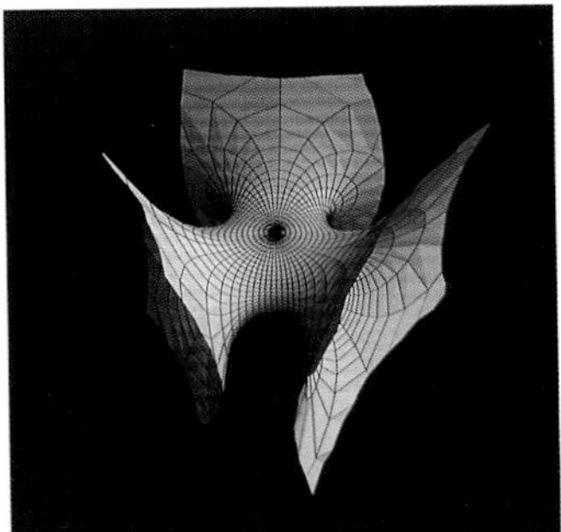

Plate 13. Michael Scroggins and Stewart Dickson, *Topological Slide* (1993). The stereoscopic pair of images above and opposite show distant views of the Jorge-Meeks Trinoid. It is possible to view each pair as a single three-dimensional image by staring at them and slightly crossing the eyes until the images merge. Above, the Jorge-Meeks Trinoid minimal surface is a sphere with three points removed from its equator. This topological transformation casts the region surrounding the missing points into funnel shapes that extend outward infinitely. The computer graphic process of discrete sampling of the parametric function results in a distortion of the funnels into angular forms. The image sequence on the left shows the rider's point of view as recorded from one video input of the stereoscopic head-mounted display. The sequence begins up near the boundary of one of the funnel-shaped extremities of the Jorge-Meeks Trinoid and progresses as the rider slides down toward the open polar region and rides along the edge. The image below is from a chromakey video composite. It shows the rider immersed in virtual space and sliding along the edge of a hole formed by the parameterization of the Jorge-Meeks Trinoid. Digital images courtesy the artists.

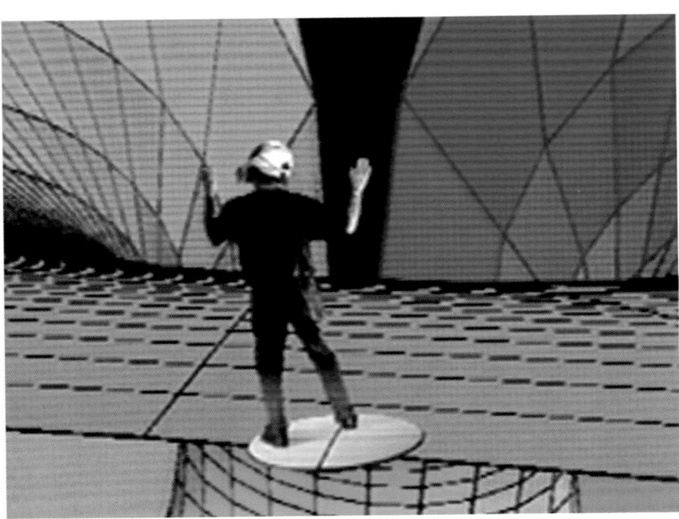

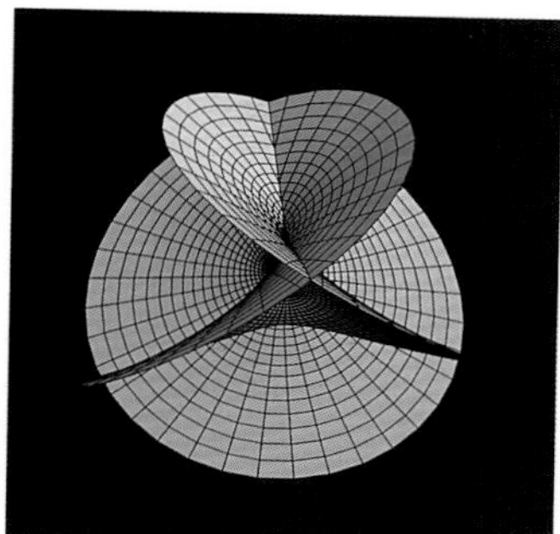

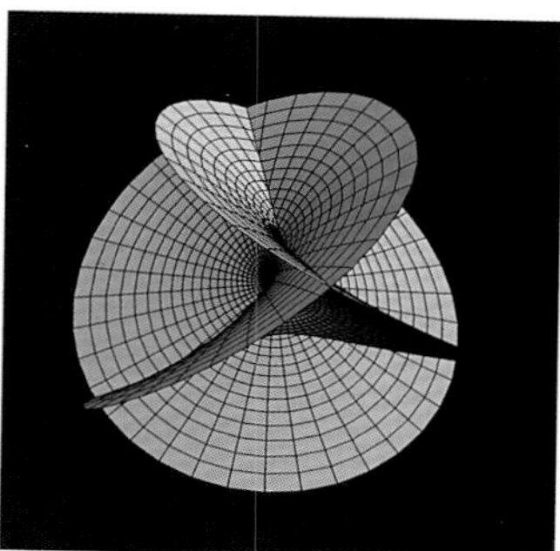

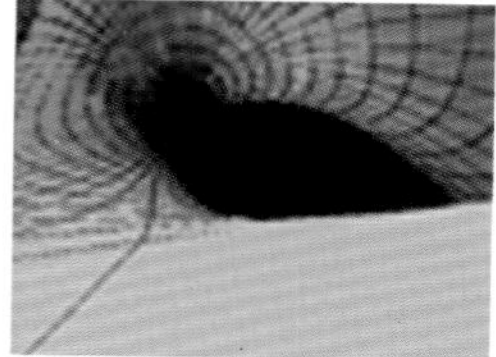

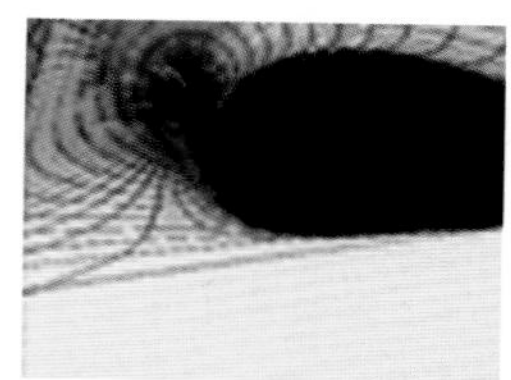

Plate 14. Michael Scroggins and Stewart Dickson, *Topological Slide* (1993). The stereoscopic pair of images above provide another distant view of the Jorge-Meeks Trinoid. This one shows Enneper's minimal surface, which is essentially a disk warped into a saddle shape that intersects itself as its edge is extended toward infinity. In the Enneper-Weierstrass parameterization used here, the edge happens to lie always on a sphere. The image sequence on the right shows the rider's point of view as recorded from one video input of the stereoscopic head-mounted display. The sequence begins as the rider slides across Enneper's surface on a region located about halfway between the center and the perimeter of the parameterization. The rider moves "under" the arching form of the "saddle," rotating his or her head in order to look up and back toward the polar center. The image below is from a chromakey video composite. It shows the rider immersed in virtual space and looking up at the overarching form of Enneper's surface as it curves away from the center point of the parameterization. Digital images courtesy the artists.

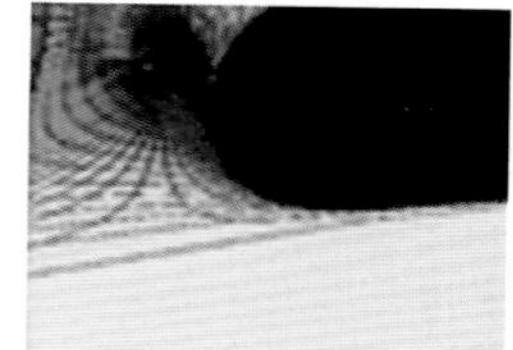

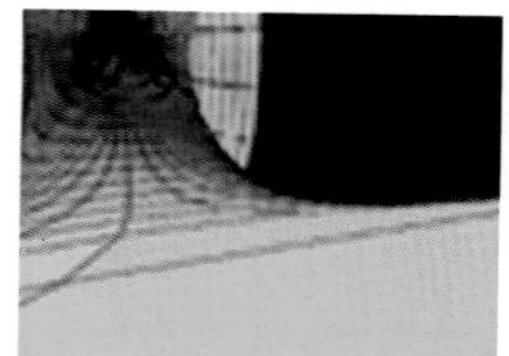

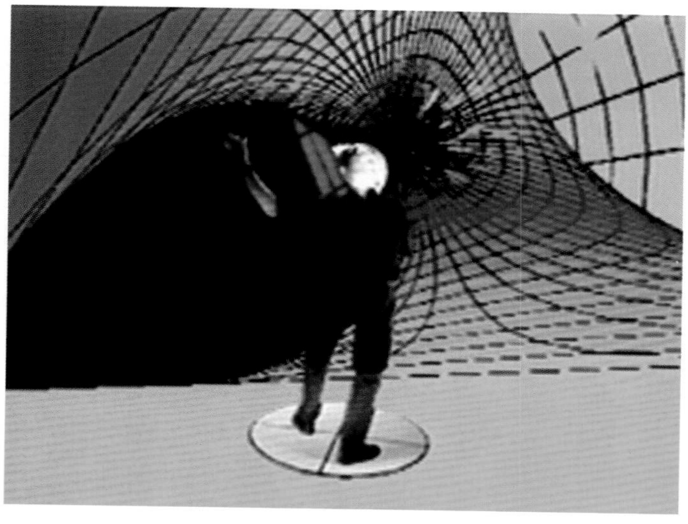

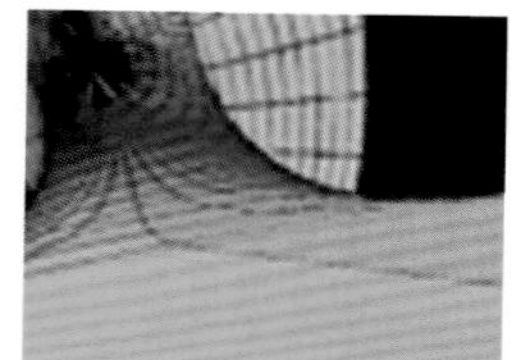

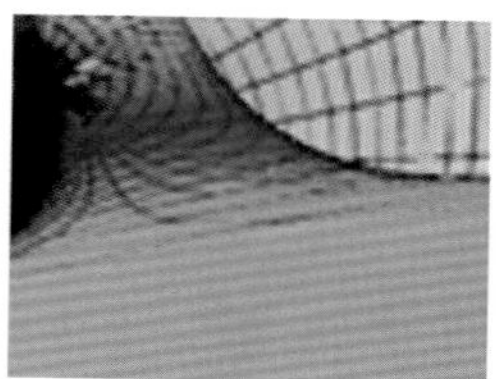

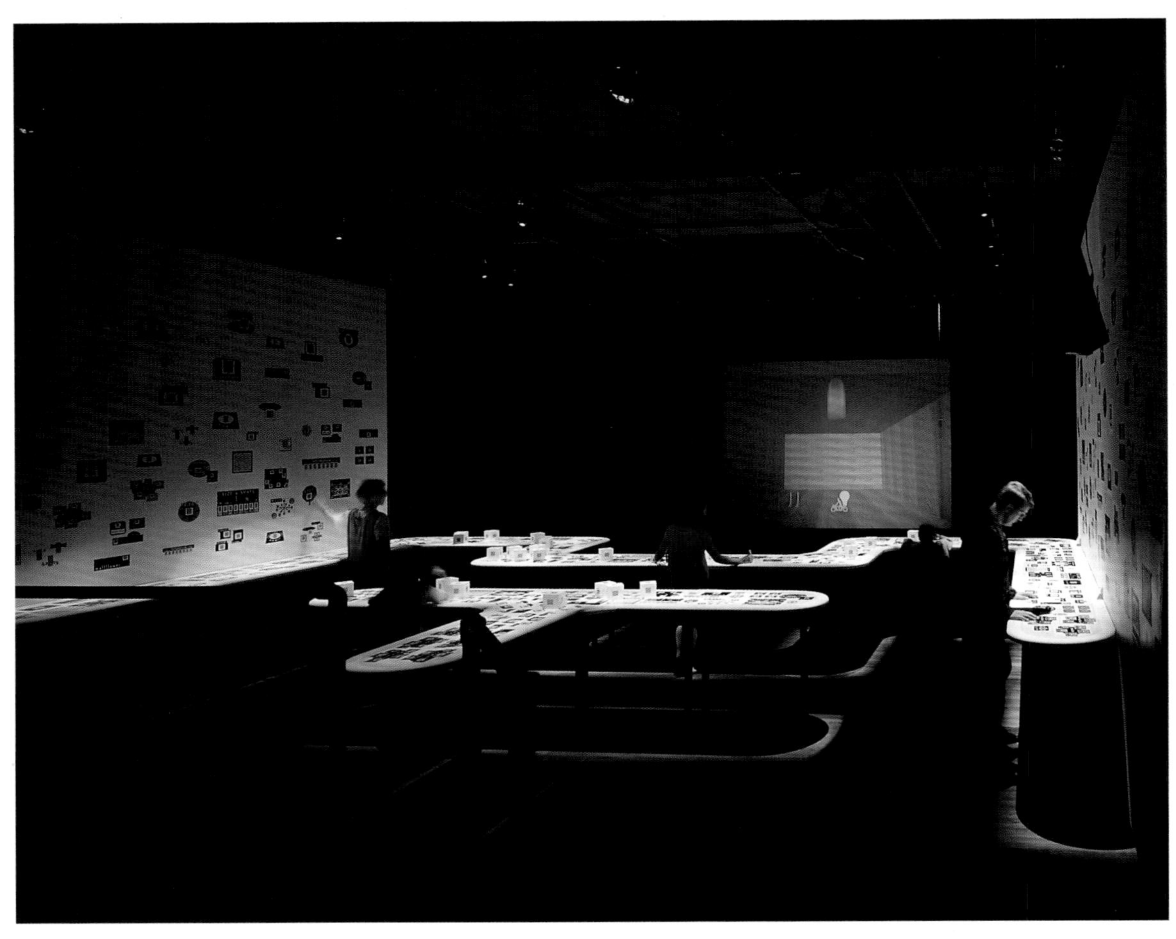

Plate 15. Perry Hoberman, *Bar Code Hotel* (1994), installation view. Photo Donald Lee.

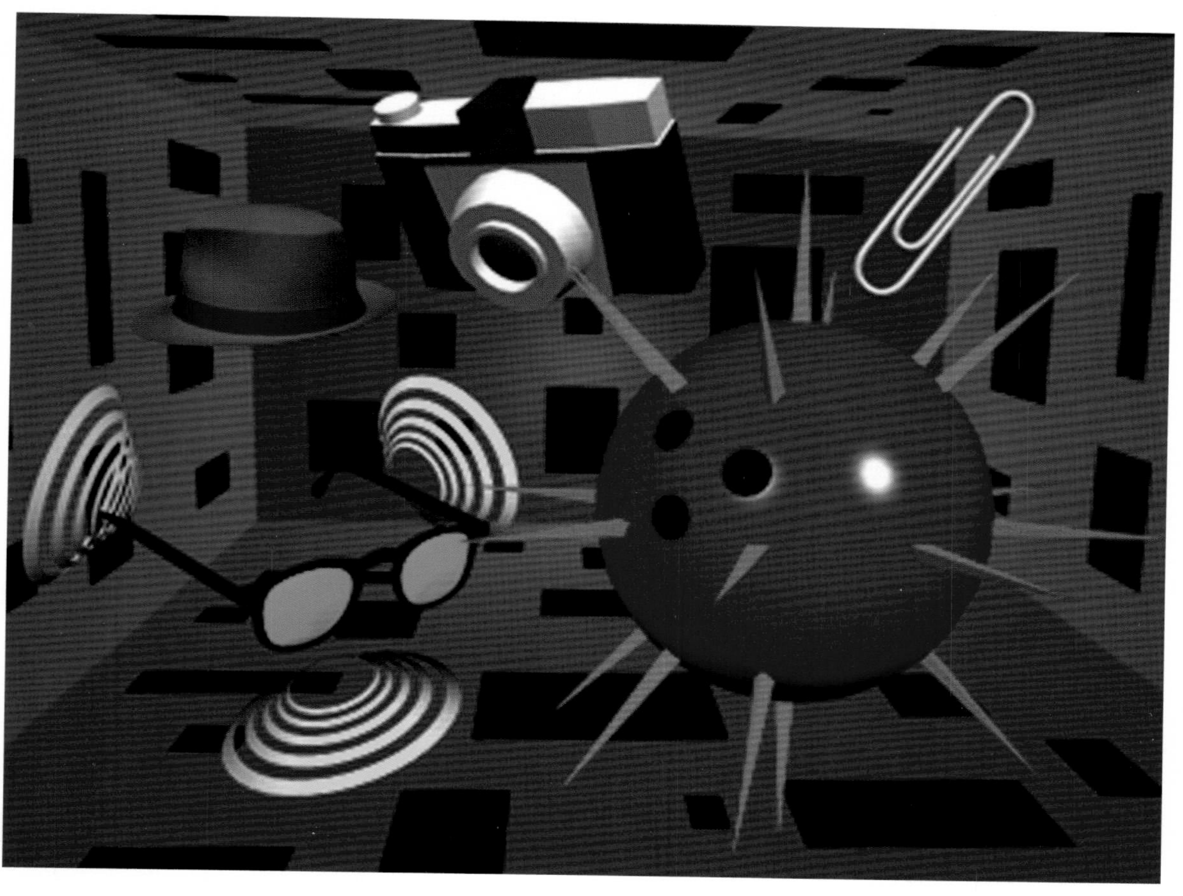

Plate 16. Perry Hoberman, *Bar Code Hotel* (1994), frame grab. Digital image Cathy McGinnis.

what became a fairly large team. It was similar to working on a film. As Michael was generating the script on this project, I became the visual director. The virtual environment became as real to me as any movie set and was as complicated to construct. Raonull Conover built the three-dimensional models, and working with him was an education in conceptualizing three-dimensional virtual space and in designing strategies for building the environment and its effects. An example of this is the setting for the wireframe rib cage in the second environment of the piece. The rib cage became a transport plane, and to create a sense of physical speed we took frame grabs from footage of industrial landscape I shot in New Jersey, processed them and mapped them onto a structure similar to a jello mold. The rib cage moved in circles inside the mold and landscape flew by.

During the year that we worked on this project there were three new generations of equipment with concomitant possibilities. Because virtual reality is such a young technology everything that happens in a virtual environment has to be programmed. The authoring tools are in an early stage of development: we had to build the piano in order to write the symphony. The creative collaboration with the programmers—John Harrison, Dorota Blaszczak, Graham Lindgren, and Glen Fraser—was for me one of the most exciting elements of the piece's development. The combination of diverse approaches produced possibilities and results that could not have been arrived at individually—was a wonderful conversation.

Michael Mackenzie

The projects sponsored by the Art and Virtual Environment Projects mark the collision between a cluster of technologies and an aesthetic purpose. This is nothing exceptional; anyone who attends an opera rehearsal will become aware of the enormous amount of accumulated technology and technique (musical scoring, instrument construction, just for starters) that goes into that form. But in opera, as in most artistic undertakings, the technology has been thoroughly integrated into *métier* and process, becoming largely invisible. The virtual environment projects had the challenge of having neither an established aesthetic nor an established process.

In about 800 B.C. an oral, recitative tradition in the northeast Mediterranean area became transformed by the impact of a technology (writing) in a new form (phonetic) that had been originally evolved for record keeping. The committing of what we now call Homeric poems to text was a profoundly creative act, an antecedent of the cultural artifact we now call literature where the spoken and written become mutually supportive and reinforcing, and only wholly understandable through the oral/text dialectic. It also created a new form of cultural remembrance.

About three hundred years later the emergence of the Greek form of theater in the same part of the world was predicated on the architectural evolution of the amphitheater and the aesthetic innovation of actors stepping from the chorus. These two factors allowed for a qualitatively new kind of intimacy and access in the presentation of the human interaction.

Such transformations in art are of course intimately connected to the changing sociopolitical conditions that engender them, and for which they, inevitably, provide commentary. Greece in the fifth century B.C., for example, was a tribal society evolving into city-states, in the case of Athens into a democratic imperial state. As such, it found itself confronted with a series of fundamental social confusions manifested as political and ethical problems. (For example, Aeschylus's first play, *The Persians,* and other later plays such as Euripides' *Medea* were about barbarians and women, that is, the other.)

Our late twentieth-century impulse is to be rather self-conscious about these connections: art has become commentary on the sociopolitical forms from which it is precipitated. The fetishistic interest in virtual reality may be based on the desire for presence, in other words, the denial of absence or invisibility. It is appropriate for the era of identity politics in which the articulate margin speaks for our generalized anxiety in its struggle (racial, gender, gender orientation, class) against negation or invisibility. VR also promises a new form of experience embodiment, namely, remembrance.

When Toni and I decided to opt for theater model as aesthetic, we were in some way opting for the obvious, and I was surprised that we were the first VR group (and the only project at Banff) to do so. I saw my role as creating a story/interactive narrative based on the specific technological opportunities and limitations of VR that also addressed the ideological context of VR (referred to above as its fetishization).

The sheer bulk of the narrative (the piece lasted 40 minutes) required that it be divided into envelopes that were sequentially loaded and unloaded into active memory on the SGI Onyx. The hiatus created by the time required to load the images was bridged by integrating an interactive laser disk player into the piece, which used video footage by Toni (in New Jersey) and myself (in Montreal) with the help of the editing skills of Jan Levis.

In the finished form, each envelope had its interactive mode appropriate to the evolving narrative at that point. The first was a dream sequence (grasping), the accumulation of reflection/memory fragments that are both haunting and evasive (Toni's figure images were exquisite) and where the text reproduces "dream as tract"—an entity that can be entered and transversed in different ways yet maintains its unity and integrity. The second envelope was narrative as happenstance experience (orientation in unfamiliar, destablizing space), provoking in turn associative memories and questions about the past for the protagonist. The third envelope develops an amnesiac second character (already glimpsed marginally in the preceding envelopes) as inversion of the primary character, experiencing memory-in-flight rather than memory-as-predatory (our first character), and whose relationship to his past (information storage) is mediated by a VR environment (mechanistic choice interactivity). The modes of interaction, developed with programmers John Harrison, Graham Lindgren, and Glen Fraser, were perhaps one of the most grueling and interesting points in the process. In one case the complexity required that we set the envelope up as best we could and run it till we hit a bud (a dead end).

The great technical support team in the program at Banff contributed to the value of the artistic experience. The key to developing the work artistically was not to construe the story in a mechanistic-interactive way, but rather to elaborate those aspects of it that dwelt on contemporary identity issues implied by what I have called the fetishization of VR: body identity, socioenvironmental identity, technology as identity, presence and absence, technology as remembrance.

The degree of success of the piece for me could be measured by the fact that participants, far from getting the experience they expected in VR, reported themselves disturbed or distressed by the experience. Good. That is an emotional journey; it worked.

Dancing with the Virtual Dervish: Virtual Bodies

Diane J. Gromala and Yacov Sharir

Diane J. Gromala

My first intention in the Art and Virtual Environments Project dealt with exploring experiential issues as they relate to notions of the body, not only as a culturally constructed notion and text but also as lived experience and material form. The aspect of materiality was especially important since discussions of the experiences attendant on virtual environments, such as disembodiment, tend to underscore an underlying subscription to the well-worn Cartesian mind/body split. As Sandy Stone reminds us, there is a real body attached, and real bodies feel pain. In my case, this is an inescapable material fact of my experience, since I daily experience chronic pain.

The result of experiencing years of chronic pain is that, like others who exist in this state, some of my perceptual faculties have become altered. My ability to hear, for example, is heightened at upper frequencies. This is experienced less often as a superhuman power than as a stressing out of perceptual faculties, with sensory input approaching uncomfortable levels. Of necessity, I have had to develop two basic strategies to deal with my various physical alterations, pain, and increased volume of perception. The first strategy was to develop an ability to reach a transcendent or disembodied state at will, related perhaps to the experience of a dervish.[1] The second strategy was to develop a sensual or erotic response to pain, embracing an enjoyment of eros and thanatos. These strategies represent less a denial of my body than a reconfigured inhabitation.

My first technologically mediated virtual experience was with my own body when, awake during surgery, I watched my own viscera being altered and manipulated on the big screen in a research hospital. In addition, electronic technologies such as X rays, magnetic resonance imaging (MRI), sonograms, and fiber-optic microcameras have enabled visualizations of material facts of my being unavailable in a technologically unmediated realm. Over the course of my life, these "studies" have outnumbered the usual snapshots and photographs one could expect to document a life. They are not, however, objective and value-free, but rather become a site where social, political, and economic forces are contested and negotiated, the consequences of which have a very direct and real effect on my material, physical existence. Another consequence of this process has often been an abrogation of the spiritual dimensions of what Elaine Scarry describes as the inexpressibility of pain.[2]

Thus, a cyborg materially as well as theoretically, I explored these issues in the creation of a virtual environment. This was an effort to reappropriate, reinhabit, reclaim, and reconstitute fragmentary representations of my body, and in turn to offer this reinscription and reconfiguration of my body as a virtual stage for both a dancer and audience interactors. This was in keeping with the collaborative group's initial themes of book, body, and architecture. It also adhered to my concern with the experiential, as I discovered that within an immersive environment one generally experiences a markedly more viscerally and kinesthetically profound response to organic than to rectilinear forms.

Literally and figuratively a body of enormous scale, the resultant virtual environment is an incomplete torso comprised of a skeletal spine, pelvis, and ribs, along with viscera of a heart, kidney, and liver. This body is programmed to remain in continuous motion, slowly undulating in sensuous movements toward decay and reformation, abstractly recalling breathing and in some cases dissolution. Although it was computationally too demanding to transform or triangulate the entirety of my MRI data into the virtual body, the three-dimensionally rendered wireframe models approximate it. These are texture-mapped or wrapped with text, as well as with X rays and abbreviated MRI data of my body. My textual meditations on the sensual and erotic related to each of the seven veils associated with dervishes.[3] In turn, each of the veils bore a relationship with a visceral component of the virtual body. These organs can be "entered" to reveal otherworldly chambers. The virtual body thus becomes an immersive, nonlinear book, a text to be read, an architecture to be inhabited.

Within the body stands another primary component, another body, video grabs of a dancer transcribed onto a plane. The dancer exists both as representations within the virtual environment and as performer in the physical performance space, connected to the virtual environment through the umbilicus of the head-mounted display.

Yacov Sharir

As a choreographer and dancer, I create work kinesthetically, emphasizing movement, space, and the body. Through the Virtual Environments project, I questioned what worlds open when, as a dancer, I consider virtual environments in relation to the body. What artistic, intellectual, kinesthetic, and emotional issues could be addressed using this technology? How could these technologies affect the creation and experience of the body by a dancer and by others?

Virtual technologies allow us to manipulate, extend, distort, and deform information as well as experience of the body. They are vehicles that enable us to extend and color work in many ways, some of which are not possible in the physical realm and/or by traditional means. They offer a way to augment and extend possibilities creatively, experientially, spatially, visually, sonically, and cognitively.

Dancing with the Virtual Dervish: Virtual Bodies is among the first virtual environment projects to synthesize immersive and interactive digitized new dance in a distributed performance environment that includes a head-mounted display, dataglove, three-dimensional sound, and interactive video projections. In performance, I create multidimensional dance while interacting with the virtual environment.

As a user/dancer, I wear the head-mounted display, which totally blocks the physical space from my vision, replacing it with a three-dimensional, stereoscopic computer simulation. As I turn and shift my focus, the simulation changes accordingly. I target my vision and/or move my hand forward, and am thus able to navigate through the simulation, birdlike. As my perception accommodates itself to the three-dimensional illusion, I accrue a sense of being in another, additional skin; I feel immersed. Yet as I experience immersion in the simulation, I simultaneously lose the sense of being grounded in physical space, I experience the sensation of disembodiment. However, my body still remains in the physical realm, constrained by the apparatus and tracking range. The sensation of disembodiment cannot be disconnected from the sensation of embodiment, physicality, groundedness. I lose the sense of which environment my body exists in. This problematizes my experience, since the sense of being grounded is usually primary to a dancer's experience. However, this also opens areas for further exploration artistically.

The experiential aspects related to the body are taken further in *Dancing with the Virtual Dervish: Virtual Bodies,* since the virtual environment I move through and within is the organic space of an enormous torso. Within the architecture of this torso, I also find digitized images of myself dancing, again distributing and multiplying my own body and experience. This mirror effect, this dancing a duet with myself, provokes a sense of heightened anxiety, caused by the doubling of my own body image. Further, my actions initiate cause-and-effect relationships in all worlds; my "body" or presence, then, is engaged with, represented by, and distributed among the many worlds. *Dancing with the Virtual Dervish: Virtual Bodies* explores concepts and experiences of the body on many levels.

The physical limitations of the tracking device, the head-mounted display, and dataglove were capitalized upon for artistic ends. Working with this new technology left us with many provocative questions. Is the nature of dance altered by this potential? Where does one locate the performance—in the performance space or the virtual environment itself? Both? Could the boundaries of the definition of dance be altered by this new technology? Could VR users be influenced to move in dancerly ways? How could agency and self be altered? How can a virtual environment and the possibilities of distributed performance redefine performance venues?

Notes

1. Dervishes are Muslim mystics who, through rituals of whirling, howling, or chanting, induce an ecstatic, transcendent state.

2. Elaine Scarry, *The Body in Pain: The Making and Unmaking of the World* (New York: Oxford University Press, 1985).

3. Desire, negligence, hypocrisy, avarice and parsimony, haste to fatigue, and separation.

Bar Code Hotel

Perry Hoberman

Perhaps there is an increasing tendency when we are working with developing technologies to downplay (or even ignore) their shortcomings. For certain technologies, such spectacular promises are made that it becomes almost easier to foresee what they could become than to see quite clearly what they are. Whether or not these promises are ever fulfilled is another question. They have led to an age in which technological paradigms shift about every half year. Almost every month seems to offer radically new media. This is undeniably useful when promoting and marketing something, but it can be misleading and counterproductive for artmaking.

Technologies are not usually created for artists. New technologies generally serve the interests of commerce, the entertainment industry, the government, and the military. Artists who want to work with technology have essentially three choices. They can be developers of new technology (that is, if they have the resources); they can be consumers of existing technology (somewhat less expensive); or they can be scavengers of old technology (really cheap). By and large, this last alternative has been my approach.

One side effect of all this relentless change is that we are left with unprecedented amounts of obsolete, outmoded, or merely passé equipment and ideas. This puts a bit of a strain on the very concept of obsolescence. What do we call technologies that are still in use but have been around so long that they become part of the background noise of modern life? And what about the technologies that never quite made it past the fad stage, either because they were poorly promoted or else because they were just plain ridiculous?

These offshoots could be described as examples of "debased technologies"—technologies that have filtered down into our lives to the point that they are impossible to avoid but easy to ignore. Debased technologies include the obsolete, the unglamorous, the ominous, the ridiculous. They are embodied in such things as security systems, slide shows, 3D movies, karaoke, random-dot stereograms, vinyl LPs, home appliances, and bar code.

In the brave new world of interactive multimedia, artists are often thought of as content providers for predefined, easily marketable formats, packages, or artifacts. In stark contrast to the idea of artist as content provider is a conception of the artwork as incorporating an apparatus that is specific, custom-made, and possibly unique. I find this second approach far more compelling, which may be why I am drawn to debased, hybrid technologies. Away from the distracting sheen of the state of the art, one can begin to investigate alternatives.

As a result, I approached this project with more of an idea of what I did not want to do than of what I did. Most of the virtual reality that I have seen requires an endless wait on an interminable line, only to briefly enter a rudimentary world in which one is a solitary inhabitant with nothing to do. Encased in gear that entirely shuts out the real world, one is immersed in the homogeneous space of computer-generated images, in which no real discontinuity of any kind is possible. I had something else in mind, at once more social, more casual, and more disruptive.

Upon arrival at the Banff Centre, one is issued an ID card with a photograph and a bar code symbol, which is passed through bar code readers in various locations for a range of purposes. One afternoon I swiped my card to pay for a meal, and was promptly informed that I had received some messages. I had never thought much about bar code before, but it struck me as strange that such entirely different functions could be conjoined. I started to research bar code technologies, and began to visualize a room in which every available surface was covered with bar code, an entirely black-and-white environment.

Bar code is the system of black and white stripes that is branded onto the surface of almost every single salable item in the western world. Introduced in the early 1970s and largely unchanged since then, bar code appears timeless, almost style-free. The Universal Product Code organizes all consumables into an exhaustive digital hierarchy that interfaces seamlessly with an ever-growing network of computer systems that can track everything from vast warehoused inventories to the minutiae of individual consumers' choices and whims.

Bar code represents an early attempt to bridge the gap between the physical world and the computer. As such, it is the forerunner of present-day attempts to allow computer comprehension of the world as it is (such as optical character recognition and artificial vision), as well as plans to embed digital information invisibly in the environment (ubiquitous computing and augmented reality).

Unlike these contemporary projects (which strive for subtlety, invisibility, a policy of nonintrusion), bar code is blatantly, unavoidably obvious. In fact, this very obviousness became its appeal to me. Bar code makes no attempt to disguise its sole function as an extension of the domain of the digital computer.

As my ideas developed, I started to think of using bar code technology as a kind of caricature of a more conventional virtual reality apparatus. Participants would be "immersed" in a computer-generated world; but this world, instead of being made up of virtual polygonal objects, would be made up of printed black and white symbols. The bar code wands were a kind of cheap substitute for a data glove. Finally, I decided to incorporate three-dimensional projection, so that participants had to don 3D glasses as a surrogate for the inevitable head-mounted display. Because I wanted to create a social, multiparticipant installation, I decided to forego the immersive first-person point of view found in most virtual reality.

These ideas coalesced into the *Bar Code Hotel,* an interactive installation for multiple participants (or guests). An entire room is covered with printed bar code symbols, creating an environment in which every surface becomes a responsive membrane, an immersive interface that can be used simultaneously by a number of people to control and respond to a projected real-time, computer-generated, stereoscopic, three-dimensional world.

Each guest who checks into *Bar Code Hotel* dons a pair of 3D glasses and picks up a bar code wand, a lightweight pen with the ability to scan and transmit printed bar code information instantaneously into the computer system. Both sound and image are highly spatial. Stereoscopic images are produced by a pair of video projectors, and localized sounds emanating from each of the objects are reproduced using a quadrophonic system (yet another debased technology). Each time a guest scans a bar code, contact is reestablished between that guest and his or her object. Between these moments of human contact, objects are on their own. This allows for a number of possible styles of interaction.

Bar codes can be scanned to modify objects' behaviors, movements, and location. Objects can expand and contract; they can breathe, tremble, jitter, or bounce. Certain bar code commands describe movement patterns, such as drift (move slowly while randomly changing direction), dodge (move quickly with sudden unpredictable changes), and wallflower (move into the nearest corner). Other bar code commands describe relations between two objects: chase (pursue nearest object), avoid (stay as far away as possible from all other objects), punch (collide with the nearest object), and merge (occupy the same space as the nearest object). Of course, the result of scanning any particular bar code will vary, based on all objects' current behavior and location.

Each object develops different capabilities and characteristics, depending on factors like age, size, and history. For instance, younger objects tend to respond quickly to bar code scans; as they age, they become more and more sluggish. Older objects begin to malfunction, flickering and short-circuiting. Finally, each object dies, entering briefly into a ghostly afterlife. (This process can be accelerated by scanning suicide.) After each object departs, a new object can be initiated.

Since any bar code can be scanned at any time, the narrative logic of *Bar Code Hotel* is strictly dependent on the decisions and whims of its guests. It can be played like a game without rules, or like a musical ensemble. Objects and their surroundings can be manipulated. It can seem to be a slow and graceful dance, or a slapstick comedy. And because the activities of *Bar Code Hotel* are affected both by its changing guests and by the autonomous behaviors of its various objects, the potential exists for the manifestation of a vast number of unpredictable and dynamic scenarios.

VR on $5 a Day

Ron Kuivila

VR on $5 a Day attempts to critique VR, the movie,[1] while exploring VR, the gizmo. The Banff realization takes the form of an interactive installation periodically interrupted by fictional advertisements. After several of the ads, the installation temporarily adopts a new form that demonstrates the product.

The Gizmo

New technology usually creates a hierarchy of access. I wanted to thwart that tendency. The installation is conceived as a virtual environment *shared* by a person in a head-mounted display ("the helmethead") and many unencumbered visitors ("the people"). To create a direct correspondence between the virtual space of the helmethead and the physical space of the people, the people's motions are digitized and embedded in the simulation. These motions appear as enormous constellations of points in the (virtual) sky; they disappear as soon as the motion stops. The helmethead appears in the simulation as a small-scale head-mounted display rendered in a homey wood grain.

There is also a flock of autonomous agents that orbit around as a group. They are attracted to the motions of people. In fact, they feed on the yellow funnels that are left by people's motions. The agents take on different forms in different states of the piece. At various times, they become head-mounted displays, mutant datagloves, and the components of a head (eyes, eyebrows, nose, lips).

The people and the helmethead have complementary abilities. The people are somewhat godlike: they are able to see both a video projection that offers a bird's-eye view of the simulation and video monitors that show the helmethead's view. Since the helmethead can only see things in close-up, she or he often gets lost and has to ask the people for directions. However, this person has total control of the close-up view and is the only participant to be immersed in the simulation.

The Critique

Commercials, the closest thing to "theory" that mass culture provides, offer a remarkably convenient form for critique. What follows are the texts of six advertisements in *VR on $5 a Day.*

Metaphor Mapping
Chastened by the failure of artificial intelligence, encouraged by our faith in human intelligence, we at Infomatique Oblique have followed the trail first blazed by the Hungarian mathematician Laszlo Funtek. His concept of the fruitful functor, as elaborated in *The Bureaucracy of Mind,* showed that game space is a dual space for the *problem set of humanity*. IO pioneered the new science of *Ludic Analysis* that Funtek foresaw.

The result? Metaphor Mapping© technology, the root of our corporate tree. Simply put, Metaphor Mapping allows any problem or task to be reexpressed as an interactive game. Once ludified, the problem is posted to our Intelligence Mine, where millions of crack game players, our Intelligence Miners, can have a go at it.

As the I-miners play, winning strategies appear and our automated Ludic Analyzer, Laszlo, collects the strategies for deludification and application.

No problem is too big or too small for Metaphor Mapping. You don't believe it? Well, listen to this! Laszlo is also IO's CEO!

Intelligence Mining

IO's intelligence mine is the place where games make gains in brains. The ever expanding variety of game in our mine has created the kinder, gentler wage slavery of current corporate culture. It's a little difficult to fit the I-mine in the mind's eye, as it spans the mediascape. So we put Laszlo and the I-team to the task, and they developed a special representation just for our guests.

But, before we proceed, let's set the record straight. Bizarre rumors have been spread that I-miners lose all sense of the consequences of their own actions and become dangerous sociopaths. These rumors are totally false. So, as you enter the I-mine, leave your prejudices behind. You can't have a minor mind if you want to meet a mind miner! And look out for light bulbs! They are signing someone's smarts for harvest!

[After the ad, the basic scene is replaced by a large aquarium tank filled with mutant datagloves.]

Personal Husbandry

No, silly, Personal Husbandry is *not* some kind of sexual separatism. PH is a way you can live high off the hog while eating low on the food chain. With our enzymatic diet supplements, you can grow your own steaks—or even spare organs—while taking only the simplest stuff out of the food chain. Become the first organot on your node, and remember . . .

[jingle] You never know where that cow has really been, so the only steak to eat is the one from your shin!!!

[hushed, golf course voice] Since the introduction of Personal Husbandry the incidence of organ abduction has declined by over 97%!

Personal Husbandry is best performed in our own "Reality Lite" sensory deprivation chambers. We don't have time for a full sitting, but here's a little example of what it's like . . .

Portfolio Personification

Contemporary financial instruments are making it harder and harder for investors to stay in contact with their financial resources. Researchers at IO have found that rapid capital redeployment with its frequent consequent, unemployment, can create grave anxieties in the *capital resources themselves.* This can lead to nonmaximal yield. IO has developed Portfolio Personification, a suite of services that help investors empathize with their capital.

If you are having problems with your portfolio, just bring it to IO. Our fiscal beauticians can work miracles with even the most complex investment strategies. After they work their magic, all you'll have to do is say "hi" and your stock's most intimate secrets will appear, exquisitely rendered, on the face of your choice.

[After the ad, the scene is replaced by a head whose component parts fly off to the center of action whenever people move.]

Safe Sense Technology with Social Typing

Life wasn't the same, after the infoduct got going. From Orange County to Soho, people clamored for ways to disconnect undesirables and reconnect to others just like themselves. It was IO's visionary leadership that led to the enactment of the Mandatory Multiphasic Personality Inventory legislation. Now, that little daily test we take keeps us aware of just where you're at. Connecting to the inventory makes it easy as pie to erase all those noncorrelates who don't fit your view. It can make the entire infoduct feel just like home.

Feeling a little wild? Want to meet some *strangers?* IO's proprietary propriety threshold control lets you loosen up and feel the difference. Set it to zero and Lord knows what you'll find!

Excerupt

OK, you helmetheads—it's time to shake that pinky. And here we go:

Index up Index down Index out Index back
Index up Index down Index out Index back
1 and 2 and 3 and 4 and
1 and 2 and 3 and 4 and
1 2 3 4 1 2 3 4

1 2 3 4 1 2 3 4
1234 1234
1234 1234
1234 1234
1234 1234

There . . . Now feel the twinge. And come back for more real soon.

Some Concluding Remarks

In some real sense, the Art and Virtual Environments Project existed more within a cultural futures market of governmental and industrial sponsorship than an art world. This had effects both subtle and dramatic both on administrative decisions and on artistic practice. For example, some participants initially viewed their projects as industrial research requiring the protection of nondisclosure agreements. This turn, while neither unreasonable nor uninteresting, is certainly problematic. The ideology underlying cultural sponsorship that intends to promote industrial growth warrants careful scrutiny.

Most of my pieces rely on inexpensive materials, my own skills and contributions begged, borrowed, and stolen from others in the shareware circuit. At Banff, I found myself placed in a disorienting but not totally disagreeable role as the manager of a small technocultural atelier. *VR on $5 a Day* could not have been completed without the able assistance and willing creativity of Raonull Conover, Glen Fraser, Graham Lindgren and Cathy McGinnis, the words of John Harrison, and the overall planning of Doug MacLeod.

Notes

1. The word movie can be used to mean a captivating rumor, as in: "There was a movie going around the festival that narcs were selling bad acid."

Placeholder

Brenda Laurel and Rachel Strickland

It depends where on earth one goes. Experiences are said to take place. One comes to know a place with all senses in concert and by virtue of the actions that one performs there, from an embodied and situated point of view. The mind, observes naturalist Barry Lopez, is a kind of projection within a person of the place that person inhabits. "Each individual undertakes to order his interior landscape according to the exterior landscape." The environment proceeds to record our presence and actions and the marks that we place there—this is a reciprocal affair.

Placeholder is the name of a research project that explored a new paradigm for narrative action in virtual environments. The geography of *Placeholder* took inspiration from three actual locations in Banff National Park: a cave, a waterfall in Johnston Canyon, and a formation of hoodoos overlooking the Bow River.

Three-dimensional videographic scene elements, spatialized sounds and voices, and simple character animation were employed to construct a composite landscape that could be visited concurrently by two physically remote participants wearing head-mounted displays. Participants were able to walk about, speak, and use both hands to touch and move virtual objects.

People's relationships with places and the creatures that inhabit them have formed the basis of many traditions and spiritual practices as well as ancient stories and myths. The graphic elements in *Placeholder* were adapted from iconography that has been inscribed upon the landscape since paleolithic times. Narrative motifs that reveal archetypal characters of landscape features and animals were selected from aboriginal tales. Four animated spirit critters—Spider, Snake, Fish, and Crow—inhabited this virtual world.

A person visiting the world could assume the character of one of these spirit animals and thereby experience aspects of its unique visual perception, its locomotion, and its voice. Thus the critters functioned as "smart costumes," altering more than the appearance of the people wearing them.

Travelers sometimes leave marks in natural places: pictograms, petroglyphs, graffiti, or trail signs, for example. In *Placeholder,* people could create voicemarks —bits of spoken narrative—that might be listened to and rearranged by subsequent visitors. The virtual landscape accumulated definition through messages and story lines that participants left along the way.

We hope that ideas we explored in *Placeholder* will foster the emergence of new forms of narrative play.

Field Recording Studies

Michael Naimark

Field Recording Studies began as an attempt to study "place" by making three-dimensional computer models based on the physical world. I wanted to retreat from making finished work by making modest studies around art and virtual environments. I did not expect to be looking back a hundred years.

Study #1: Panoramic Tiling and Guerrilla Technology

This first study was intended to be fast and cheap. (In fact, I flew into Banff exactly two weeks before I had to fly out with something to exhibit in Siggraph's *Guerrilla Technology* show.) The concept was to make one contiguous panorama by "tiling" together slightly overlapping still images. These images were collected with a simple 8mm video camera on a tripod, using traditional surveying tools such as compass and level.

The site I selected was a viewpoint above the Banff hoodoos overlooking the Bow Valley and Rundle Mountain, particularly spectacular just after dawn.

Back at the Banff Centre John Harrison had prepared software for hand-positioning the forty-two still images we digitized from the video for a single panorama. No human should have to do this, but we had no choice. The images were meticulously lined up so the edges matched, and the result was a single panoramic dome.

Study #2: Projection Sculpting and *Placeholder*

The virtual dome was still composed of flat two-dimensional images. The next step was to shape the dome to the actual contours of the landscape, by hand if necessary. Conceptually, this process, of taking two-dimensional images and making three-dimensional shapes, is exactly what a sculptor does when working from photos. Humans are remarkably good at filling in the missing information based on our everyday knowledge.

Shortly after I completed Study #1, Brenda Laurel and Rachel Strickland approached me about collaborating on their project, *Placeholder*. John Harrison believed that such "projection sculpting" could be possible with his new software and the Banff Centre's new SGI Onyx. I had been wagering with my colleagues at Interval Research that if I had a slide projector, a slide of a human face, and a giant mound of mashed potatoes, I could hand-shape the mound to create a credible (if not accurate) three-dimensional face projection. I accepted Brenda and Rachel's offer.

The first site they selected for projection sculpting was Johnston Falls. Rachel had a short video of the falls and its surroundings. Several frames were digitized (to provide a sense of motion) and turned over to Cathy McGinnis, who was to hand-shape the flat image. Several (actually many) attempts were made at shaping, as well as both shaping and tiling from another site, Troll Falls. But the most credible was the first one made by Cathy. It became the setting for one of the three *Placeholder* virtual worlds.

Study #3: 3D Moviemapping and the Hundred-Year-Old Connection—*See Banff!*

At this point I gave up on computer technology. We were using the world's most powerful graphics computer and it kept choking. The industry, and art, is built around making computer models from scratch rather than from cameras. The result has been a bias toward making fantasy worlds, imaginary places, and scientific visualizations rather than representations based on the actual world, however abstract.

I reverted to film. Wonderful medium, film: simple, portable, rugged. Rich vibrant colors, subtle detail. Working with my research colleagues back at Interval and with artist Mark Pauline, we built a modest camera rig out of two old 16mm cameras and a three-wheeled "super-jogger" baby carriage. The cameras were mounted side by side for stereographic 3D and had very wide-angle lenses, similar to those in virtual reality goggles. The cameras could only shoot one frame at a time, and could be triggered either by an intervalometer for time lapses or by an encoder on one of the carriage wheels for moviemaps. The idea was to capture imagery that "looked like VR" enough to encourage further investigation.

The camera rig could be disassembled to fit in a car, and reassembled and ready within minutes. With Gilles Tassé, a filmmaker with extensive location experience in the area, I spent six weeks filming studies around Banff and rural Alberta. We filmed over a hundred sequences.

Our biggest struggle was not technological (thankfully) but artistic. On the one hand, the sites in the Banff area are monumental in their grace and beauty. Some are sacred. On the other hand, watching the tourists at these sites told a different story. Buses and buses pull in and out of parking lots seemingly in the middle of nowhere. Tourists get out—cameras in hand—for twenty minutes, then get back on the buses, off to another site. Gilles and I agreed that the strength of the footage would lie in juxtaposing these two conflicting messages.

Banff has a curious history as a tourist area, which dates back a hundred years. It has experienced a longstanding controversy over tourism, ecology, and growth. For example, automobiles were banned from the Banff National Park in 1904, a ban not completely lifted until 1917. Ecological controversies continue today.

In a real sense, wilderness tourism in Canada was invented by the Canadian Pacific Railway, which built a chain of luxury hotels to subsidize the construction of the railway west at the turn of the century. Also at that time, visual representation technology was exploding. Stereoscopes—stereo photographs mounted on card stock and viewed with simple viewers—were popular then. "Around the world . . . without leaving your home/Just like being there" was the slogan of a popular stereoscope series. They proved to be a very high-quality and effective means of viewing our footage in three dimensions.

Pre-cinema public exhibitions were also proliferating. The most popular featured the Edison kinetoscope, which made its public debut in April 1894. One and a half years later, in December 1895, the Lumière brothers publicly exhibited projected film, and cinema as we know it was born.

The kinetoscope became a transitional symbol during a turbulent era in the media arts. Making one out of our Banff footage seemed like the right thing to do. *See Banff!* is the artwork that resulted.

Dancing with the Virtual Dervish: Worlds in Progress

Marcos Novak

What could be more secret than a vision? Difficult to attain even for the one who has it, it is painfully impossible to share. Consider the image of a "whirling dervish"—a Sufi mystic blind to the world but spinning in a secret vision. We can see the person, we can see the spinning, but we cannot enter the mental universe within which she dances. Now, compare the image of the dervish to that of a person donning the late twentieth century's version of the mystic's robe: the head-mounted display, the dataglove, the tangle of wires. Confined to the narrow radius of sensor-reach, joined to the ceiling by an umbilical cord connecting brain to computer, eyes blind to the world, this spinning consciousness is also lost in a vision. The parallel is evident and strong, but there is a key difference: this vision is constructed, and can thus be shared.

Virtual worldmaking is like writing books that very few people will ever encounter, let alone read. The limitations to access imposed by the technology and its costs are still severe. Even those who have experienced the many chambers of *Dancing with the Virtual Dervish* directly have only explored a small part of its territories—they have only read the book's cover. This brief exposition itself only addresses a few of the issues and motivations that guide the ongoing creation of the Dervish's worlds.

Dancing with the Virtual Dervish, in its present disincarnation, consists of a growing number of "world chambers" that form a fully connected lattice, a rhizome. All the worlds lead to all the other worlds. There is no narrative hierarchy, no purposive spurring, no beginning or end, no closure. No genealogy, no teleology, no eschatology: only ontology. Once inside these worlds, in the midst of a tangle of technology, you are left to your own devices.

Five worlds exist: teleworld, stageworld, bodyworld, virtualworld, cyberworld; nested within each of these, a host of others, some worked out, some still to come, worlds in progress. They form a continuum from the externally distant, to the nearby, to the intimately near, to the simulacrum, to the internally distant. *Dancing with the Virtual Dervish: Worlds in Progress* aims at creating a work that spans these various worlds and allows events in each to affect the outcome of events in others. Performances at remote sites are to influence the direction of performances on local stages; events onstage are to alter the course of events within the virtual worlds; the activities of one or many participants are to propagate inward and outward to the stage and world, creating strange feedback loops that constitute a living, global artwork.

At this writing, the most developed aspect of the project is that of the cyberspace chambers themselves. The number and makeup of these worlds continue to grow with each showing of the work. A sampling of the initial list of worlds includes the following: a world in which there is a small heart within the center of an architectural maze, and the heart is an eye and a portal, and the whole sky is filled with a constellation of hundreds of indecipherable smiles, and the smiles are calm as long as you stay near the heart, but they spin ever more wildly the farther you get from it; another world in which an impossibly slender floating city—grown by L-systems, algorithms that simulate the growth of plants—is found within another, vast heart, and the texture of that heart is made of bones, and if one gets lost in the chambers of that heart one may never find the way out, unless one carries Ariadne's thread; another world that is cavernous, crystalline, made of Sufi patterns, within which arrays of shapes alternate between algorithmic dances and the swimming patterns of schools of artificial subaquatic life. These worlds are connected by portals floating in space that disappear once you fly through them, as if the doors you walk through could vanish behind you and reemerge elsewhere.

The first history this work is concerned with is the history of space. Every artwork projects a virtual environment, and Coleridge's willing suspension of disbelief is required in order for the spectator to become immersed in a work's particular universe. Virtual environments replace the spectator with the inhabitant, and consequently the terms are reversed: the spectator does not suspend disbelief, but rather the spectator-become-inhabitant is held suspended within the arms of unbelievable space. Primary among the concerns of this project, then, has been the concern of speaking through space, through liquid, virtual space itself.

Space is a prime symbol of the aspirations of a people, permeating architecture, music, and mathematics. The emergence of a new spatial conception announces the passage of one culture into oblivion and the arrival of another. Cyberspace is unlike any other understanding of space ever conceived; it is the hallmark of a new era, as different from the one that is ending as the Gothic was from the classic. As such, it demarcates an ending and a beginning.

This project affiliates itself with a forward-looking, though always remembering, present. Combining music, plasticity, number, and space, into a new medium I call "ArchiMusic," the Dervish project seeks to speak the first utterances of what will become a new multimode of expression.

The second history is that of the body, "the body in pain," to borrow Elaine Scarry's phrase, in the struggle against embodiment that characterizes our religions, laws, moralities, and technologies. Within the memory of the Dervish the human figure appears in vividly juxtaposed images of torture and beauty, dance and cruelty, inescapable carnal reminders of that which we are tempted to leave behind as we enter cyberspace. To raise the tension further, even as we explore worlds of extreme abstraction, we are brought back to our bodies: within these alien worlds the most mathematical of objects, the most analytical of behaviors, are at the same time the most erotic and sensuous presences one can encounter. Navigating through continuously curved mathematical landscapes, guided by the subtlest modulations of the orientation of the dataglove, one has the distinct sense of gently caressing a lover's body.

A major concern of this work is how to have the virtual environments themselves be the main expressive medium. Beyond representation, purpose, or plot, I use the virtual environments themselves as chords of space, intonations of place. The explicit meanings one may project onto or into the Dervish are secondary to the meanings that are evoked by the sense and intonation of the virtual chambers themselves, the pulses within them, the dilation of time, and the multitimbral disparities of their imagery.

One chamber stands out: it is, to the best of my knowledge, the world's first immersive experience of the fourth dimension. By this I do not mean time as the fourth dimension, but a fourth spatial dimension. A series of proto-architectonic four-dimensional objects rotate (in the fourth dimension, of course) around a vestige of the Cartesian coordinate system. All their vertices have four coordinates, all that would appear to us as planes are, topologically, cubes, all that would be cubes, hypercubes. Projected into three-space, their shadows are three-dimensional objects that perform a complex and graceful transformational dance. Walls advance gently toward the viewer, pass right through, and continue. With time, one learns to read the shapes, and when they are aligned correctly, one can actually see recognizable proto-architectonic figures.

All of the chambers of the Dervish are designed along a continuum ranging from lesser to greater physical impossibility. This chamber is at the far end of that range: not only is it impossible to build outside cyberspace, even after experiencing its three-dimensional projections we cannot claim to know its four-dimensional reality.

The music in *Dancing with the Virtual Dervish* is an early effort at what I call "navigable music," music conceived as a slowly evolving landscape or datascape, composed as a multidimensional "surface" across which the viewer travels. What is heard at any given time is the outcome of the viewer's trajectory, fast or slow, interesting or uninteresting in direct proportion to the energy and curiosity invested in exploring the various chambers. All aspects of the sound are contingent upon the actions of the viewer, his or her position, velocity, attitude, trajectory, history. Moreover, the music is not passive, but actively alters the turns of events within the chambers: as the compositional algorithms track the viewer and make decisions about the sonic events, they also trigger visual events in the form of semiautonomous dancing forms within the environment. The media components of the Virtual Dervish are early investigations into the media of dis/embodiment and their use in the perpetual construction of a liquid self.

The creation of virtual environments announces a new instrument, what I will call the "esoscope," the instrument with which to examine the inner worlds of the self. To share the visions of the Dervish is to peer within the psyche.

Topological Slide

Michael Scroggins and Stewart Dickson

Michael Scroggins

The original grant application to the Art and Virtual Environments Project proposed the use of a tiltable platform as a kinesthetic interface for navigating upon topological surfaces. It went on to describe the project thus:

> The "rider" will wear a head-mounted display enabling an interactive wide-angle stereo view of a three-dimensional space. The space will consist of a model of a topological surface to which the platform is bound and upon which it is free to slide. The "rider" may traverse the model's surface by leaning in the direction in which she desires to move. The amount of lean in a given direction will determine the rate of sliding.

This simple concept was the heart of the proposal, and there was no conscious attempt to deal directly with the many interesting and complex conceptual issues surrounding VR. The *Topological Slide* was conceived as a direct sensual experience of surface with overt handles for the intellect located in the mathematical concepts integral to the surface's formation.

In the early stages of the project, I tried to discourage any associations with surfing, preferring instead to invoke the more general model of sliding. My attempts to avoid the surfing tag were doomed, however, in light of the prevalent use of surfing as a metaphor. No one seemed to be speaking of "channel skating" or "internet skiing"!

Nonetheless, the germinal idea for the *Topological Slide* did arise in an idle daydream of a VR surfing simulator. A constant tubular wave generator could allow the visual illusion of riding inside the most intense part of a wave. Getting tubed can be a transcendent experience as the level of concentration required to remain inside the tube can propel one into a state of mind where there is no past and no future, just a lucid present. Unfortunately, the surfing simulator idea unavoidably degenerated into the annoying notion of a kind of VR karaoke where one could replace Frankie and Annette as surf heroes.

An attempt to recuperate some aspect of the surfing model led me to consider utilizing the continuous looping flow of a Möbius strip as a substitute for a simulated wave. Reflecting on the topological properties of the Möbius strip led to consideration of the Klein bottle and then on to the wide range of minimal surfaces that Stewart Dickson had brought to my attention years earlier.

My modernist indoctrination gave weight to the idea that traversing mathematical objects was much more interesting than riding a weak replica of an ocean wave and that experiencing the abstract-made-concrete was clearly a more valuable process than experiencing the actual-made-virtual. There is a long and rich history to the linkage of art and mathematics, and the *Topological Slide* project may be viewed from several positions in this stream.[1] One way of thinking of the project is to consider the topological surfaces as a priori objects presented as art. Jung remarked that numbers might be seen to be as much discovered as invented by man,[2] and Duchamp established that the found object could take on a power equivalent to that of the crafted art object.[3]

A contributing factor to my personal fascination with the *Topological Slide* is my mathematical naïveté. In addition to the sensual delight of navigating the surfaces in a very physical way, I wanted to gain an understanding of the principles by which those surfaces are formed and the underlying concepts that make them interesting to mathematicians. In the course of dealing with these objects, my curiosity has been aroused enough to desire further education in mathematics. I have been forced to come to terms with how little I understand of non-Euclidean geometries and thus how limited my concepts of space have been. It is my hope that this project will also stimulate others—particularly children—to expand their mathematical understanding.

I asked my old friend and colleague Stewart Dickson to collaborate with me on the project because of his deep interest in visual mathematics and his experience designing topological object databases. I was responsible for the design and fabrication of the tilting platform, and as the person who conceived and initiated the project my primary role was to coordinate its overall realization. Stewart provided the code for the topological objects and was responsible for writing and implementing the original platform interface routines. Subsequent programming necessary to realize the project was performed by Graham Lindgren, John Harrison, and Glen Fraser. Further work in optimizing the *Topological Slide* code and implementing edge control algorithms was performed by Sean Halliday.

STEWART DICKSON

The construction of minimal surfaces by contemporary mathematicians like David Hoffman and William Meeks is an exercise in obtaining a specific topology with the geometry constrained to be the purest possible expression of the form. The breakthrough work of Hoffman and Meeks used the computer to make pictures of the forms they were studying. The pictures showed them aspects of the objects they could not deduce from the equations alone. They began doing a kind of experimental mathematics that revolutionized their field of work. They began designing mathematical systems that were driven by a visually derived structural model. They then had a higher-level point of view from which they could advance their formal rigor. Documentation of this work can be found at <www.gang.umass.edu>.

The forms we have chosen to use in the *Topological Slide* are artifacts of the historical legacy of cyberspace.

The general problem of placing the rider on a mathematically specified surface is fairly complex. First the (x,y,z) coordinate in three-space on the mathematical surface must be calculated as a function of the (u,v) coordinate in parametric space of the rider's location. This location is the cumulative effect of the two acceleration vectors specified by the platform tilt axes and a constant friction or drag coefficient. The rider must then be oriented in a natural manner to the surface at the correct three-dimensional address.

The parametric surface I am using in this piece is generally represented by a polynomial of fourth degree or higher. *Costa's Genus One, Three-Ended Minimal Surface* is specified as the real parts of three integrals of expressions involving the Weierstrass elliptic p-function over regions in the complex plane. In general, solutions to an arbitrary polynomial equation of degree five or more cannot necessarily be written as algebraic expressions. For equations of degree exactly five, the solutions can in principle be written in a complicated way in terms of elliptic functions; for higher-degree equations, even this is not possible. This is a fundamental difficulty in the state of contemporary mathematics.[4] The surfaces we are exploring are indeed on the fringe of human knowledge.

In short, these calculations are nontrivial. I rely on *Mathematica* to obtain graphical representations of the surfaces. Even if I could derive C-language implementations of the calculations, it would not be advantageous to calculate them on the fly; therefore, we used precomputed CGI-style geometry description records to represent the surfaces.

I had originally intended that the *Topological Slide* contain an object that is in metamorphosis from its simplest form (disk, sphere, or torus, for example) into "full bloom." This action was intended to better illustrate the nature of the shape as well as to add dynamism to the riding experience.

Another class of surfaces we intended to illustrate was the three-dimensional representation of an entity that is defined in a higher-dimensional space. I have prepared pairs of immersions in three-space of the *Real Projective Plane* that metamorphose one into another: the Veronese surface that metamorphoses into the Steiner Roman surface and the Apery cylindrical parameterization of the Roman surface that metamorphoses into the Werner Boy surface. I also have two different (orthogonal) parameterizations of the Klein bottle that transform one into the other. Further development time would have allowed me to realize these goals.

Notes

1. An excellent collection of essays on the subject can be found in Michelle Emmer's *The Visual Mind: Art and Mathematics* (Cambridge, Massachusetts: MIT Press, 1993).

2. Carl Jung, *The Collected Works of C. G. Jung,* vol. 8, "The Structure and Dynamics of the Psyche" (Princeton: Princeton University Press, 1969), 457.

3. The ironic inversion inherent in this reading can be discerned in Calvin Tomkin's *The Bride and the Bachelors* (London: Weidenfeld and Nicolson, 1965), 26.

4. Paraphrased from Stephen Wolfram, *Mathematica: A System for Doing Mathematics by Computer,* 2d ed. (Reading, Mass.: Addison-Wesley, 1990).

Inherent Rights, Vision Rights

Lawrence Paul Yuxweluptun

I am an artist of two cultural circumstances. Since I happen to be a native, I paint the color of life, I see all this land, in a native way. I was born to see it this way. I find my life painting history from a native point of view. I was born in the time of the nuclear weapon, which was invented even before I was born. I am documenting neo-times. I approach my work and have always approached it taking my social and ideological responsibility as an artist first, as they relate to this time in history. I cannot celebrate or feel any national allegiance to the Canadian flag while such racist legislation as the Indian Act remains in force: the system native people are governed under is the despotism of white self-interest. Because of this, a lot of my pieces are historical. You cannot hide the real history or even the censorship of native history, a colonial syndrome. You can hide Department of Indian Affairs documents from the time of Confederation, but you cannot hide my paintings. They are there for all people to see.

I paint on a reservation, to feel this perspective on life—the racist segregation of Canadian history, the so-called "Indian problem." I will paint you, O Canada, for all your atrocities face my paint, for they are part of my life. Try to read the symbols for what they are.

I am concerned with the colonial mentality that is directly responsible for the killing of wolves, buffalo, whales, grizzly bears, and migratory birds, to the point that some species are now extinct. It is behind the depletion of fish stocks on the west and east coasts, acid rain, nuclear waste, land fills, smog, the greenhouse effect, the emission of methane gas, mining tailings, endangered animals, pollution of freshwater reservoirs, toxic wastes of all kinds, oil spills, uranium mining, nuclear testing. The next step toward decolonization of First Nations must be recognition by the provincial governments of our sovereign indigenous government. As sovereign caretakers of the land, our forebears were always the protectors of the biosphere.

My work is very different from traditional artwork. How do you paint a land claim? You can't carve a totem pole that has a beer bottle on it. I find myself coming back to the land. Is it necessary to totally butcher all of this land? The grizzly bear has never signed away his land, why on earth would I, or a fish, or a bird? To slowly kill my ancestral land? All the money in the bank cannot buy or magically bring back a dead biosystem. I paint this for what it is—a very toxic land base. This is what my ancestral motherland is becoming. Painting is a form of political activism, a way to exercise my inherent right, my right to authority, my freedom. This is real freedom for me. I am proud these days, I have self-dignity in my art when I paint this world. I see environmental "schmuck," so I paint "schmuck"-art in all its toxicological bliss. I can speak out in my paintings even without the recognition of self-government. I dance around in a longhouse, I dance around fires in turn, like my forebears have done since time immemorial. I am a preservationist, continuing my heritage.

Inherent Rights, Vision Rights is a virtual reality project I have been involved with for some time now. I approach it from the aspect of the fear others have of native people. Not understanding our spirit world. In it, the longhouse is a given space in time which I use to show a religious concept, to physically bring people into contact with a native worshiping aspect of life, praying Indians—a way to bring others close to my heart so they can understand my belief system. What it is like being in a possessed state feeling rhythmic sounds in a longhouse, feeling sounds go through one's own self, feeling a spirit inside you. I have been a Blackface dancer eighteen years now, a masked dancer, a *Sxqyxwey* dancer since I was fourteen. I have been able to draw from these native experiences, combining them with western world experiences and technology to make my work. Employing technology that in the past has been used against native people. I created *Inherent Rights, Vision Rights* to show people what is happening to me spiritually. Always, I create art to communicate with others, to let other cultures see things for themselves. To show my world, *Indian* world, to show that we do have a spirit, a place to go to, so people will understand who I am as a West Coast native person.

Although this particular "virtual reality" is very primitive at this point of evolution—it has limits in visual resolution—the capabilities of the technology will change in years to come. At the moment, the piece consists of a white man's mask, the "helmet," as it is called by the computer program. A screen goes over your eyes covering part of your face, an electronic mask with an electronic-motion hand. You start to experience a new art form. As you look into the mask, the screen shows you a piece of artwork, computer-electronically stimulated into color. Sound can be brought in at the same time. I think this first mask will end up in museums just like other masks! Very primitive, with numbers on them, and the date they were made.

I'm also a British Columbia artist painting from a British Columbia position. This is different from some other native points of view. But ozone is still my problem. The ozone layer is a rare form of oxygen found nine to thirty miles above this planet's surface.[1] The chlorofluorocarbons (CFCs) used in spray cans, refrigerators, and air conditioners have gradually released chlorine atoms that have destroyed the ozone above the Antarctic and the North Pole. Ten percent of the ozone here will vanish by the decade's end. In response to this threat the industrial nations have agreed to phase out CFCs and other ozone-depleting chemicals by the year 2000. It will take a hundred-year ban to bring ozone levels back to pre-1985 levels before the hole in the ozone layer can be fixed.

I don't have any rights in this country, so I paint the ozone as a problem.

Other environmental hazards are destroying and disrupting the ecosystems of First Nations' lands. They include groundwater evaporation because of big cities and clearcut logging, both in Canada and the United States. The province of British Columbia alone is sixty years behind in silviculture; it is cutting trees faster than it is replacing them.

There is a need for a moratorium on the number of logging truck licenses granted to the forestry industry. There is a need overall for an environmental audit to track companies' ecological performance on an annual basis and to inform the public, as well as for information-sharing on green environmental policies and programs at the governmental and community levels.

At the same time, the world's population is growing and will double in the next thirty-eight years. It is imperative to apply first world environmental standards to third world factories, but global treaties to extend economic and technical assistance to developing nations are needed to help them help themselves to develop ecologically safe practices. We need these treaties to integrate environmental criteria into economic practices, and to provide management, technical, and financial resources to tackle global problems. The economic efficiency of green policies must be improved, because the toxicological time bomb has already gone off, threatening all life forms with health hazards and, ultimately, toxic death.

Good luck, O Civilization, for every hundred years I will ask you to drink the water with me, my Brother and Sister, and I will wait for you to take the first drink.

Notes

1. Figures quoted in this essay are from Ed Magnuson et al., "Poisoning of America," *Time* (14 October 1985), 64–80; and Melanie Menagh, "Report Card on the Planet," *Omni* (July 1992), 30–40.

Artwork Production Credits

The Art and Virtual Environments Projects was produced by the Computer Applications and Research Program of the Banff Centre for the Arts. The artworks in this book were made possible through the tremendous dedication of a small team of professionals. John Harrison, the senior research analyst on the project, and Dorota Blaszczak, the sound designer, worked under impossible deadlines to create the works shown here. Raonull Conover and Cathy McGinnis, computer associates, again and again proved their artistry as they constructed the models and graphics used in these pieces. Glen Fraser and Graham Lindgren, virtual reality programmers, constantly demonstrated their virtuosity in solving difficult programming tasks. And Douglas Smith, computer support specialist, supplied expertise and support to each of the projects in his usual patient and accommodating manner. Far more than any overarching planning or piece of equipment, the skill and hard work of these people made these projects possible. Moving well beyond the roles of technicians, this group quickly came to be viewed as essential collaborators by all of the artists involved in the project.

None of these projects would have happened without months of planning and organization. Michael Century, then Director of Inter-program Development at the Banff Centre and now at CITI (Centre for Information Technologies Innovation), was instrumental in orchestrating the funding of the project. Garry Beirne, then Head of Computer Media at the Centre and now at the University of Toronto, was essential to the initial planning of the project.

Special mention should also be made of the Virtual Reality Lab in the Department of Computing Science at the University of Alberta. Its software, MRToolKit, was the foundation on which these projects were based. Dr. Mark Green, director of the lab, and Christopher Shaw worked closely with the Banff Centre throughout the project, and their contribution was invaluable. At the Banff Centre, Rick Bidlack was instrumental in the development of the audio software used in the project.

Individual areas of responsibility for each of the projects and the contributions of others are gratefully acknowledged as follows:

Objects of Ritual

Conceived and directed by Will Bauer and Steve Gibson
Script: Will Bauer
Sampling, sequencing, and sound editing: Steve Gibson
Gesture and media system: Will Bauer
SGI models and slide manipulation: Raonull Conover, Cathy McGinnis
Star maker narration: Rebecca Chua
NeXT sound processing: Dorota Blaszczak
Senior research analyst: John Harrison
Virtual reality programmers: Glen Fraser, Graham Lindgren
Audio associate: Jennifer Lewis
Executive producer: Douglas MacLeod
Installation: Gordon Tait
Audio: Chris Seignitz

Objects of Ritual
© 1994 Will Bauer and Steve Gibson

Archeology of a Mother Tongue

Conceived and directed by Toni Dove and Michael Mackenzie
Programmers: Glen Fraser, John Harrison, Graham Lindgren
Sound designer: Dorota Blaszczak
3D designer: Raonull Conover
Computer graphics assistant: Douglas Smith
Video and video post-production: Edith Labbé, Jan Levis, Jody Polowick, Dani Singer, Einar Westerlund
Additional sound design: Rick Bidlack, Jan Levis, Jennifer Lewis, Dana McCurdy, Antonio Oliart, Stephanie Rogers
Performers: Arielle Amenoff, Gianna Corbisiero, J. J. Gifford, Sylvie Gilbert, Mari Kimura, Claire Pentecost
Character development: Judy Nylon
Executive producer: Douglas MacLeod

The artists would like to acknowledge the assistance of Janet Anderson, Michael Century, Sara Diamond, Vern Hume, Douglas MacLeod, Don Stein, Piranesi, The Canada Council, National Arts Centre of Canada, and Art Matters, Inc., New York

Archeology of a Mother Tongue
© 1994 Toni Dove and Michael Mackenzie

Dancing with the Virtual Dervish: Virtual Bodies

Conceived and directed by Diane J. Gromala and Yacov Sharir
Text, X rays, body parts, and dances: Diane J. Gromala, Yacov Sharir
Algorithmics, architectural environments: Marcos Novak
Senior research analyst: John Harrison
Sound designer: Dorota Blaszczak

Composer: Russell Pinkston
Virtual reality programmers: Glen Fraser, Graham Lindgren, Chris Shaw
Computer associate: Raonull Conover
Audio associate: Jennifer Lewis
Computer support specialist: Douglas Smith
Videographer: Jay Ashcraft
Graphic art assistant: Jana Wilson
Executive producer: Douglas MacLeod

Dancing with the Virtual Dervish: Virtual Bodies
© 1994 Diane J. Gromala and Yacov Sharir

Bar Code Hotel

Conceived and directed by Perry Hoberman
Senior research analyst: John Harrison
Sound designers: Dorota Blaszczak, Glen Fraser
Virtual reality programmers: Glen Fraser, Graham Lindgren
3D Designer and project assistant: Cathy McGinnis
Audio associate: Jennifer Lewis
Computer support specialist: Douglas Smith
Installation: Mimmo Maiolo
Installation assistants: John Baturin, Michael Cameron, Glenda Montgomery
Executive producer: Douglas MacLeod

The artist would like to acknowledge the assistance of Steve Gibson, Ron Kuivila, Douglas Smith, and Angela Wyman.

Bar Code Hotel
© 1994 Perry Hoberman

VR on $5 a Day

Conceived and directed by Ron Kuivila
Senior research analyst: John Harrison
Sound designer: Dorota Blaszczak
Virtual reality programmers: Glen Fraser, Graham Lindgren
3D designers: Raonull Conover, Cathy McGinnis
Audio associate: Jennifer Lewis
Recording engineer: Jan Levis
Computer support specialist: Douglas Smith
Voices: Perry Hoberman, Ann Carlson
Executive producer: Douglas MacLeod

VR on $5 a Day
© 1994 Ron Kuivila

Placeholder

Produced and directed by Brenda Laurel and Rachel Strickland
Architectural design: Rachel Strickland
Theatrical design: Brenda Laurel
Software design and engineering: John Harrison
Audio design and engineering: Dorota Blaszczak
Camera-originated scenes: produced in collaboration with Michael Naimark, Field Recording Study #2
Audio engineering and field recording: Rob Tow
Graphic design: Russell Zeidner
Mythology and folklore: Lucinda deLorimier
Landscape modeling: Raonull Conover, Cathy McGinnis
Grippee design and engineering: Steve Saunders
Additional software engineering: Glen Fraser, Graham Lindgren, Rob Tow, Sean White

Improvisation and voices: Precipice Theatre Society, Colin Funk, director
Production crew: Jennifer Lewis, Mark Richardson, Chris Segnitz, Kerry Stauffer, Gordon Tait, Gilles Tassé, Luke Van Dyk
3D audio technology: Crystal River Engineering
Executive producers: Douglas MacLeod, Banff Centre for the Arts; David Liddle, Interval Research Corporation

The artists would like to acknowledge the assistance of Janet Anderson, Scott Foster, Rebecca Fuson, Noël Hirst, Carol Holmes, Stephanie Rogers, Diane Schiano, Chris Shaw, Andrew Singer, and Douglas Smith

Placeholder
© 1993 Brenda Laurel and Rachel Strickland

Field Recording Studies

Conceived and directed by Michael Naimark

Executive producers: Douglas MacLeod, Banff Centre for the Arts; David Liddle, Interval Research Corporation

Principal collaborators
Film production: Gilles Tassé, Banff Centre
Video post-production: Charles "Bud" Lassiter, Interval Research
Software and optics: Christoph Dohrmann, Interval Research
Interface electronics: Robert Alkire, Interval Research
Cabinetry and mechanics: Robert Adams, Interval Research
Technical assistance: Scott Walters, Interval Research
Special cabinetry: Andy Hope, Piranha Design, San Francisco
Special mechanics: Mark Pauline, Survival Research Labs, San Francisco

Additional collaborators
Canada:
Banff Centre: Lorne Falk, John Harrison, Cathy McGinnis, Mary Anne Moser
And: Roy Cross, Banff; James Martin, Canadian National Parks Service, Banff; Mireille Perron, Alberta College of Art, Calgary; Banff Lifts Ltd., Banff; Capital Film Lab, Edmonton; Columbia Icefields Tours, Banff; Studio Post, Edmonton; Whyte Museum, Banff
United States:
Interval Research: Aviv Bergman, Lee Felsenstein, David Gessel, Noël Hirst, Jason Lewis, Carol Moran, Steve Saunders, Chris Seguine, Richard Shoup, Andrew Singer, Rachel Strickland, Bill Verplank, Cedric Whigham, John Woodfill, Ramin Zabih
And: Thomas Cockrill, Byron Antiques, Byron, CA; Judith Donath, MIT Media Lab, Cambridge; Lisa Goldman, Interactive Media Festival, San Francisco; L. H. Lassiter, M.D., Miller Eye Center, Chattanooga; Daniel Zelinsky, Musée Mécanique, San Francisco; Allied Diner Film and Video, San Francisco; Exploratorium, San Francisco; Monaco Film Labs, San Francisco; San Francisco Art Institute; Western Images, San Francisco

See Banff!
© 1994 Michael Naimark

Dancing with the Virtual Dervish: Worlds in Progress

Conceived and directed by Marcos Novak
Concept, algorithmics, and environments: Marcos Novak
Text, X rays, body parts, and dances: Diane Gromala, Yacov Sharir
Senior research analyst: John Harrison
Sound designer: Dorota Blaszczak
Virtual reality programmers: Glen Fraser, Graham Lindgren
3D designers: Raonull Conover, Cathy McGinnis
Audio associate: Jennifer Lewis

Technical support: Darryl Kaminski
Computer support specialist: Douglas Smith
Navigable music: developed in collaboration with Rick Bidlack
Executive producer: Douglas MacLeod

The artist would like to acknowledge the assistance of Dr. Mark Green, Christopher Shaw, and the Virtual Reality Lab of the Department of Computing Science at the University of Alberta

Dancing with the Virtual Dervish: Worlds in Progress
© 1994 Marcos Novak

TOPOLOGICAL SLIDE

Conceived and directed by Michael Scroggins
Programmer, objects and interface: Stewart Dickson
Senior research analyst: John Harrison
Sound designer: Dorota Blaszczak
Virtual reality programmers: Glen Fraser, Sean Halliday, Graham Lindgren
Computer associates: Raonull Conover, Cathy McGinnis
Audio associate: Jennifer Lewis
Computer support specialist: Douglas Smith
Software for sound processing on the NeXT/IRCAMM workstation: created by Rick Bidlack
Executive producer: Douglas MacLeod

The artists would like to acknowledge the advice and assistance provided by Dr. Mark Green, Christopher Shaw, and the Virtual Reality Lab of the Department of Computing Science at the University of Alberta

Topological Slide
© 1994 Michael Scroggins

Inherent Rights, Vision Rights

Conceived and directed by Lawrence Paul Yuxweluptun
Imaging and design: Lawrence Paul Yuxweluptun
Senior research analyst: John Harrison
Sound designer: Dorota Blaszczak
Modeling: Douglas MacLeod
Computer support specialist: Douglas Smith
Computer associates: Raonull Conover, Cathy McGuiness
Installation: Mimmo Maiolo
Technical support: Chris Seignitz
Executive producer: Douglas MacLeod

Inherent Rights, Vision Rights
© 1992 Lawrence Paul Yuxweluptun

Art and Virtual Environmental Project Credits

The production technical staff at the Banff Centre have been essential to the project as a whole. In particular the contributions of Gordon Tait, Kerry Stauffer, Chris Seignitz, Lucas Van Dyk, and Ben West are gratefully acknowledged. We are also thankful for the additional assistance of the Audio and Television/ Video programs as well as Technical Services at the Banff Centre for the Arts.

Valuable administrative support was provided by administrative staff of the Media Arts Department, including: manager, Dan Thorburn; coordinators, Janet Anderson and Maureen Lee; administrative assistants, Colin Funk and Judy Golumbia; and secretaries, Inge Gaida, Claire Honeyman, and Erin Michie.

The Art and Virtual Environments Symposium was organized by Douglas MacLeod (chair), Sara Diamond, Kevin Elliot, Mary Anne Moser, and Dan Thorburn, with the assistance of: coordinator, Janet Anderson; technical director, Vern Hume; conference assistants, Judy Golumbia and Erin Michie; and technical services manager, Ben West.

The Art and Virtual Environments Project is funded by the Department of Canadian Heritage (formerly the Department of Communications) and CITI (Centre for Information Technologies Innovation). The Banff Centre is grateful for additional support from Silicon Graphics Inc., Alias Research, the Computer Science Department at the University of Alberta, Apple Canada, the Intel Corp., and Autodesk Inc. In particular the work of Martin Tuori (then at Alias Research), Connie Griffiths and Dale Wieber of Silicon Graphics Canada, Michael Ede of Apple Canada, Kevin Teixiera of the Intel Corp., and Christopher Allis of Autodesk must be acknowledged.

Technical Information

The virtual environment artworks represented in this book were created using a variety of systems. Texture maps were created on Macintosh computers running Adobe PhotoShop. Soundscapes were developed on a Macintosh IIci running Sample Cell, Audio Media, and MAX and on NeXT/IRCAMM workstations (and Convolotrons in the case of *Placeholder*). Programming of the environments was done using MRToolKit from the University of Alberta and Performer from Silicon Graphics, and models were constructed using Alias Studio, all running on two Silicon Graphics VGX 310s and an ONYX Reality Rack with a Reality Engine2 and a MultiChannel Option.

Contributors

Cameron Bailey is a Toronto-based writer and programmer of film and video whose work has appeared in *The Village Voice, CineAction!, Border/Lines,* and *Screen.* He is a regular contributor to *Now* magazine and a programmer for the Toronto International Film Festival.

Will Bauer is an artist and engineer working with "integrated" (as opposed to "multi") media. He is the inventor of the Gesture And Media System (GAMS), which is now in use by artists around the world as a means for producing integrated-media works of art. He collaborated with Knowbotic Research on their Golden Nica–winning piece *S.M.D.K.* and, most recently, has worked with Rafael Lozano-Hemmer on the telepresence piece *The Trace.*

Stewart Dickson is a sculptor, computer programmer and consultant, and president of MathArt Incorporated, an Internet on-line service at URL http://www.mathart.com.

Toni Dove is a performance/installation artist who works with electronic media. Her work has been exhibited extensively, adapted for broadcast, and developed for publication. She lives in New York.

Frances Dyson is a writer and multimedia artist who has explored sound theory and practice for many years. She teaches in the faculty of creative arts at the University of Wollongong in Australia.

Steve Gibson is a composer and media artist who holds a doctoral degree in music composition. He currently serves as multimedia editor of *CTheory*. His releases include *Spasm,* a compact disk created in collaboration with Arthur Kroker.

Diane J. Gromala is an artist, designer, writer, and director of the New Media Research Lab at the University of Washington in Seattle. Her work has been published and exhibited in Europe, the Middle East, and throughout North America.

N. Katherine Hayles, a professor of English at the University of California, Los Angeles, has written extensively on the literature and science of the twentieth century.

Perry Hoberman is an installation and performance artist who works with a variety of media, ranging from totally obsolete to state-of-the-art. He has performed and exhibited extensively throughout the United States and in Europe. He is currently art director at Telepresence Research.

Erkki Huhtamo is a writer, researcher, and curator. He is currently professor of media studies at the University of Lapland, Rovaniemi, Finland. He has published extensively on media art and audiovisual culture. He has also curated several international media art exhibitions.

Ron Kuivila is an artist living in Connecticut and New York. He composes music and designs sound installations that revolve around unusual homemade and home-modified electronic instruments. He teaches experimental music at Wesleyan University.

Brenda Laurel is a writer, producer, researcher, and interface designer who currently works at Interval Research Corporation. She is editor of *The Art of Human-Computer Interface Design* and author of *Computers as Theater.*

Michael Mackenzie's plays have been produced in North America and Europe, and have been published in English and French. His work on technology policy has been published in North America, Latin America, and Europe, and variously translated into French, Spanish, and Portuguese. He is a *concepteur collaborator* of the Ex Machina group.

Rob Milthorp is an interdisciplinary artist, writer, and academic dean at the Alberta College of Art and Design in Calgary, Alberta. His work explores masculinity in contemporary culture.

Margaret Morse teaches film, video theory, and criticism at the University of California, Santa Cruz, and has written and lectured extensively on media art.

Michael Naimark is an artist and researcher who has exhibited internationally and taught throughout the United States. He is currently a researcher at Interval Research Corporation and is on the editorial boards of *Presence,* a journal on teleoperators and virtual environments, and *Leonardo Electronic Almanac.*

Marcos Novak is an architect, artist, composer, and theorist investigating actual, virtual, and hybrid intelligent environments. He originated the study of liquid architectures in cyberspace, and is founding director of the RealityLab and the Advanced Design Research Program at the School of Architecture at the University of Texas at Austin.

Jeanne Randolph has authored several books on contemporary art and lectured widely on technology, rhetoric, and psychoanalysis. She is a psychiatrist at The Toronto Hospital and an instructor at the Ontario College of Art.

Avital Ronell is associate professor of comparative literature at the University of California, Berkeley. She is the author of *The Telephone Book* and *Crack Wars.*

Michael Scroggins is an artist who has been working with real-time videographic animation systems for over three decades. He teaches both video and computer graphic animation at California Institute of the Arts.

Yacov Sharir is a performer, choreographer, and founder of the American Deaf Dance Company and the Sharir Dance Company, the resident professional dance company of the University of Texas at Austin's College of Fine Arts.

Allucquère Rosanne Stone teaches in the Advanced Communications Technologies Laboratory in the Department of Radio, Television and Film at the University of Texas at Austin. She has written and lectured extensively on cyberspace, and is the author of *The War of Desire and Technology at the Close of the Mechanical Age*.

Rachel Strickland is an architect, filmmaker, and interactive multimedia designer. She served as consulting videographer to Apple Computer's Vivarium Program for five years and currently is a researcher at Interval Research Corporation.

Nell Tenhaaf is an artist and writer currently living in Pittsburgh, where she is visiting assistant professor at Carnegie Mellon University. She is coauthor of *The Virtual Seminar on the Bioapparatus* and is represented by Galerie Samuel Lallouz in Montreal.

Loretta Todd is a Vancouver-based independent film and video artist and writer. She has been active in developing aboriginal film and video production and has produced numerous films at the National Film Board of Canada.

Lawrence Paul Yuxweluptun is an artist whose work has been exhibited in Canada and the United States. His work incorporates his concerns as a Cowichan-Okanagan about contemporary society. He lives in Fort St. James, British Columbia.

Index

Page numbers in *italics* refer to full-page illustrations

Page numbers in **boldface** refer to artists' statements